The Skeleton

The Skeleton:

Fantastic Framework

By Kathy E. Goldberg
and the Editors of U.S.News Books

U.S.NEWS BOOKS   Washington, D.C.

U.S.NEWS BOOKS

THE HUMAN BODY
The Skeleton:
Fantastic Framework

Editor/Publisher
Roy B. Pinchot

Series Editor
Judith Gersten

Picture Editor: Leah Bendavid-Val

Book Design: David M. Seager

Art Director: Jack Lanza

Art Coordinators
Irwin Glusker, Kristen Reilly

Staff Writers
Christopher West Davis,
Kathy E. Goldberg, Karen Jensen,
Michael Kitch, Charles R. Miller,
Doug M. Podolsky, Matthew J. Schudel,
Robert D. Selim, Edward O. Welles, Jr.

Director of Text Research: William Rust

Chief Researcher: Bruce A. Lewenstein

Text Researchers
Susana Barañano, Barbara L. Buchman,
Heléne Goldberg, Michael C. McCarthy,
E. Cameron Ritchie, Ann S. Rosoff,
Loraine S. Suskind

Picture Researchers
Jean Shapiro Cantú, Gregory A. Johnson,
Ronald M. Davis, Leora Kahn,
David Ross, Lynne Russillo,
JoAnn Tooley

Technical Illustration Layout
Esperance Shatarah

Art Staff
Raymond J. Ferry, Martha Anne Scheele

Director of Production: Harold F. Chevalier

Production Coordinator: Diane B. Freed

Production Assistant: Mary Ann Haas

Production Staff
Carol Bashara, Ina Bloomberg,
Barbara M. Clark, Glenna Mickelson,
Sharon Turner

Quality Control Director: Joseph Postilion

Director of Sales: James Brady

Business Planning: Robert Licht

Fulfillment Director: Debra Hasday Fanshel

Fulfillment Assistant: Diane Childress

Cover Design: Moonink Communications

Cover Art: Paul Giovanopoulos

Series Consultants

Donald M. Engelman is Molecular Biophysicist and Biochemist at Yale University and a guest Biophysicist at the Brookhaven National Laboratory in New York. A specialist in biological structure, Dr. Engelman has published research in American and European journals. From 1976 to 1980, he was chairman of the Molecular Biology Study Section at the National Institutes of Health.

Stanley Joel Reiser is Associate Professor of Medical History at Harvard Medical School and codirector of the Kennedy Interfaculty Program in Medical Ethics at the University. He is the author of *Medicine and the Reign of Technology* and coeditor of *Ethics in Medicine: Historical Perspectives and Contemporary Concerns*.

Harold C. Slavkin, Professor of Biochemistry at the University of Southern California, directs the Graduate Program in Craniofacial Biology and also serves as Chief of the Laboratory for Developmental Biology in the University's Gerontology Center. His research on the genetic basis of congenital defects of the head and neck has been widely published.

Lewis Thomas is Chancellor of the Memorial Sloan-Kettering Cancer Center in New York City. A member of the National Academy of Sciences, Dr. Thomas has served on advisory councils of the National Institutes of Health. He has written *The Medusa and the Snail* and *The Lives of a Cell*, which received the 1974 National Book Award in Arts and Letters.

Consultants for **The Skeleton**

Victor H. Frankel is Director of Orthopedic Surgery at the Hospital for Joint Diseases Orthopedic Institute and Professor of orthopedic surgery at the Mount Sinai School of Medicine in New York City. He has served as special consultant to the National Institutes of Health and as adviser to Congress and the Food and Drug Administration. Formerly Chairman of the Department of Orthopedics at the University of Washington in Seattle, Dr. Frankel has contributed to a variety of scientific journals and has authored *Basic Biomechanics of the Skeletal System*.

Stanford A. Lavine is an orthopedic surgeon at Sibley Memorial Hospital in Washington, D.C. A specialist in athletic injuries, Dr. Lavine is physician to the Washington Redskins, the Washington Bullets and the University of Maryland football and basketball teams. He is a founding member of the American Orthopedic Society of Sports Medicine and serves on the National Athletic Trainers Association advisory board.

Ashley Montagu, an anatomist and anthropologist, is well known for his examination of human social and biological evolution. His books, which number more than forty, probe a broad range of topics including genetics, anatomy and physiology. Among his works are *Growing Young, Touching, The Nature of Human Aggression* and *The Natural Superiority of Women*. Currently a visiting lecturer at Princeton University, Dr. Montagu has also taught at Harvard University and the University of California at Santa Barbara.

Picture Consultants

Amram Cohen is General Surgery Resident at the Walter Reed Army Medical Center in Washington, D.C.

Richard G. Kessel, Professor of Zoology at the University of Iowa, studies cells, tissues and organs with scanning and transmission electron microscopy instruments. He is coauthor of two books on electron microscopy.

**U.S.News Books, a division of
U.S.News & World Report, Inc.**

**Library of Congress
Cataloging in Publication Data**

Goldberg, Kathy E.
    The skeleton.

    (The Human body)
    Includes index.
    1. Skeleton. I. U.S.News Books. II. Title.
III. Series.
QM101.G63   611'.71   81–23098
                                      AACR2
ISBN 0–89193–605–X
ISBN 0–89193–635–1 (leatherbound)
ISBN 0–89193–665–3 (school ed.)

20 19 18 17 16 15 14 13 12 11
10   9   8   7   6   5   4   3   2   1

Contents

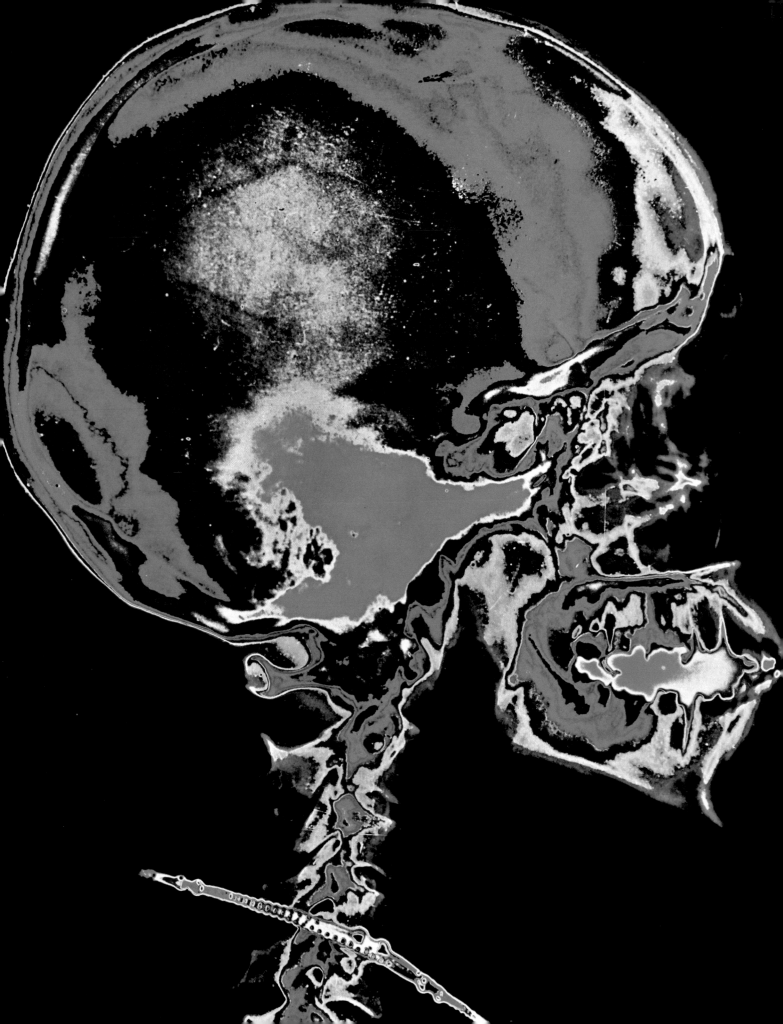

Introduction:

A Lively Legacy

"The design of a temple depends on symmetry," wrote Roman architect and engineer Marcus Vitruvius. To find such proportion, he further advised, study geometry's shrine — the human body. Arms extended, legs apart, a man simultaneously touches a square's four corners and describes a circle's perfect arc.

In the bones beneath, man finds more than the beauty of simple mathematic form. The skeleton is a measure as full and diverse as life itself. Built with the strength of an oak, it can also bend with a sapling's ease. It shelters the organs, supports the body and, bound by muscle, bestows the grace of movement. A relic that long outlives the flesh it carries, bone nonetheless meets the moment's needs. Ever building and breaking down, this dynamic tissue forms in proportion to the task at hand. The bones in a ballerina's feet, a sculptor's hands or a bricklayer's arm gain mass and shape in response to the stresses their varied pursuits impose.

In the hollow-eyed gaze of a skull man sees his mortal self. He is a vertebrate, a creature reliant on a sturdy internal frame centered by a prominent spine. Bone is his being's cornerstone. When broken, it heals with bone, not scar — a trait that sets it apart from nearly all other body tissues.

Mending through life, bone transcends death, marking the path man has traveled over time. Anthropologists retread that path, their eyes to the earth. They wait for a stir of wind or fall of rain to raise fresh fragments of bone — and awareness — to light. This search for such skeletal traces is as much an act of faith as it is of science. It mirrors in scope and ambition the task of modern medical inquiry. To fathom the mysteries of bone, scientists now painstakingly probe to crack its complex cellular codes. Their findings suggest that the skeleton, long a fund of symbol and meaning, stands as an ever-increasing store of knowledge.

Keyed by color to reflect skull tissue densities, an X-ray unveils a modern masterpiece. Man's skeletal substance underlies his versatile nature. It assures him both a living form and a lasting presence.

Chapter 1

An Ageless Form

Bards of the body, bones linger on, mute testimony to both the mortality of man and the persistence of his kind. Plutarch, Greek storyteller of the first century A.D., recorded how the Egyptians escorted a skeleton to their festivities. Seated amid the living, the skeleton urged "guests to remember that what it is now, they soon shall be." By its stark anonymity, the skeleton reminded the company, each so unique in life, that death makes all alike. Even as the spare guest dampened festive spirits, Plutarch tells us that it admonished others at the table against making their short lives bitter by courting evil. Timeless symbol of death, the skeleton at the feast prompted reflection on how to live.

From the warnings on lethal potions to the mockery of Halloween costumes, the skeleton is a poignant reminder of death. Cutthroat corsairs sailed under skull and crossbones and the dreaded Nazi SS stitched a leering white death's-head on their uniforms. When the Four Horsemen of the Apocalypse rode roughshod across medieval Europe, artists commonly counted their toll in skeletons. Used in myth and magic to render people deaf, dumb and blind, as though under death's shadow, bones were once common tools in the kits of medieval burglars. In Ruthenia, today part of Russia, burglars put bones to inventive use, believing them to be endowed with magical powers. A thief would take the marrow from a shinbone and fill it with tallow. Holding this macabre candle aloft, he circled a house three times in the hope of putting his victims into a deathly slumber. More talented thieves carved flutes from leg bones, seeking to mesmerize their marks with lullabies. In parts of Eastern Europe, burglars of coarser mettle simply threw a bone on the threshold, believing its mere presence could stupefy victims. Like the grim reaper himself, the skeleton gathers a rich harvest in the symbol, myth and magic of death.

Sitting atop the winged wheel of time, a skull dominates a mosaic from Pompeii. Like a moth circling a flame, man endlessly flirts with this sinister symbol of death, reminding himself of the infinite line of his forebears and the brief ecstasy of his life.

9

Roused by the word of God, dry bones littering the valley floor come to life before Ezekiel's eyes. Gaunt skeletons rise, stretch and limber — drawing breath from the four winds — to form the ranks of the house of Israel. Gustave Doré's nineteenth-century engraving captures Ezekiel's vision of divine will and power. The theme of life springing from dry bones resounds in the customs and folklore of people in many lands around the world.

Yet nearly everywhere this sinister remnant of life is a thread in the skein of remembered time, wound by imagination and belief, binding generation to generation. Both primitive myth and formal religion frequently endow bones with spirit or vitality surviving the deceased. The peoples of many cultures around the world carefully tend the bones of dead relatives, cherishing them as sacred relics or even talking with them as though they still lived, felt and thought. For three years of mourning, widows among the Carrier Indians of the Pacific Northwest carry with them the burned bones of their husbands. In Australia, some tribes of aborigines sever the arm bone, where spirit dwells, and bury the rest of the skeleton. Then, with elaborate ceremony, they break the arm bone, freeing the imprisoned spirit without risk either to living or dead.

New Flesh From Bones

The theme of new life springing from dry bones resounds in the myth and folklore of many lands. The Modoc of northern California tell of a tribe of skeletons in the Land of the Dead, any one of whom can be returned to the Land of the Living by four touches from a relative. But if the relative looks back before passing between the two lands, the enlivened skeleton at once turns into a heap of bones. Several biblical passages speak of bones as the source of renewed life. In Ezekiel's vision, the Lord sets him down in a valley strewn with bones, commanding him to prophesy that He will restore them to life. Ezekiel obeys and, filled with awe, watches bones assemble, sinews knit, flesh appear and skin unfold as the house of Israel fills the valley. Rabbinical tradition places the source of life in the "luz," a bone of unknown designation but perhaps an allusion to the sacrum, a sacred bone near the base of the spine. From the luz, as from a seed, can grow a resurrected body. The Church of Rome has long venerated the bones of saints and martyrs, believing they hold a measure of the grace which blessed the living. Rarely, a people so dreads the dead that they do not preserve their bones. An African tribe, the Ogowe, beat the corpse, breaking every bone, to quell the spirit and prevent it from enticing others into its fellowship of death. But in

10

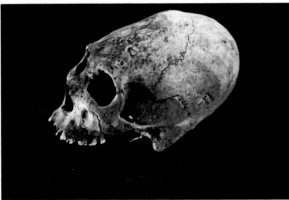

Deformed from birth, Akhenaton's skull was naturally elongated, but his queen, Nefertiti, and their daughters were shown sharing his singular profile. The royal art of Egypt set a fashion for artificially lengthened skulls which swept around the Mediterranean and through Africa. Skull shaping was also widespread among the Indians of the Americas. The skull at right, lengthened by bandaging, was found in Bolivia near Lake Titicaca.

A Chinook father cradles his child in a device designed to mold its head after the image of his own. Drawing the thongs tighter presses the board above the child's head down across the brow, lending it a marked slope.

funeral rites the world around, the overwhelming majority of cultures have preserved and venerated the bones of the dead, as though heeding Shakespeare who, in his epitaph, warned:

> Good friend, for Jesus' sake forbear
> To dig the dust enclosèd here;
> Blest be the man that spares these stones,
> And curst be he that moves my bones.

The living skeleton is not always granted the same peace and respect reserved for the dead. In most times and places people have wrenched bones into aberrant shapes, seeking beauties alien to nature. The most apparent and widespread tampering with natural proportions has been in the shaping of the skull. A solid shell, the skull can be molded without changing its capacity or damaging its contents, though the process itself may be painful. Shaping begins at birth when the child's head is soft and malleable.

The origins of the practice are puzzling. A fashion for elongated heads arose in the art of Egypt during the reign of Akhenaton, husband to Nefertiti, in the fourteenth century B.C. A heretic who introduced sun worship, Akhenaton was much celebrated in the art of his time. His skull, recovered from his tomb, was markedly elongated by nature, not artifice. Yet sculptures, carvings and frescoes portray Akhenaton, his queen and their daughters all with elongated heads. The king's appearance was almost surely unique, but royalty set an artistic standard.

In ancient Egypt the fashion was confined to royal art. But the Minoans of Crete were so impressed that they copied the fashion in anatomy. The earliest skulls known to be shaped by man are from the Minoan Bronze Age, which began thirty-five centuries before Christ and lasted another twenty-five. Whether the custom spread from one source or arose independently in several places may never be known. But it did become widespread, occurring in Asia, Europe, Africa, Oceania and the Americas.

Despite the time and space over which skull shaping has ranged, its results, methods and purposes betray considerable similarities. Elongated skulls, formed by wrapping the head with bandages, were commonplace throughout much of

Africa until recent times. Ancient Peru, a rich and varied gallery of skull sculpture, also yields many skulls lengthened by similar techniques. But in the Americas, as well as Oceania, flattening and broadening the head with planks and boards was more usual. The Chinook tribe of the Pacific Northwest stretched the infant on a plank and fixed a board obliquely over the forehead. After a year the child had a flattened crown, broadened head and receding forehead. In the Caucasus, a mountainous region between the Caspian and Black seas in present-day Russia, the child's forehead was flattened by a cap lined with pads and bound with wraps. Polynesians sought to flatten the skull by placing heavy stones around the head, one at the top and one on each side, as the baby lay in its cradle.

In Europe, skulls were shaped chiefly for aesthetic reasons. Tightly fitted caps, worn by Dutch boys until they were seven or eight and women throughout their lives, pressed the forehead down and angled it back. Bandeaus worn by French children flattened and lengthened their heads in such different degrees as to reveal their province of origin. Elsewhere, a sculpted skull marked nobility, serving as a physical sign of social rank. Although seldom if ever practiced nowadays, the custom of shaping the skull has its counterpart in contemporary hair styling.

Beauty Crippled

Less widespread but more harmful was the binding of women's feet in China. For over a thousand years, until the practice was outlawed and abandoned half a century ago, women of rank were all but maimed to satisfy the peculiar fetish for the lotus foot. At age five or six, a girl's feet were bound to force the heel and forefoot, except for the big toe, as close together as possible. The four remaining toes were curled underneath the ball of the foot. As she got older, the bonds were tightened, stunting the natural growth of her feet. Once formed, the lotus foot was marked by a steeply curved instep crowning a deep cleft beneath the arch. Prized for its erotic appeal, the lotus foot was required of all but the most common women, and especially of prostitutes and paramours. With dainty feet and mincing step,

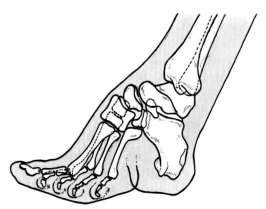

13

Bound fast forever, flint and bone speak starkly of scores surely settled. Loosed by some ancient archer, this arrow of forgotten fortune found its mark, piercing the breastbone far enough to strike the heart and kill.

these women were crippled by their culture's sense of beauty. The lotus foot has its modern counterpart in high-heeled shoes.

If man has long molded and twisted his bones, he has for even longer sought to mend and heal them. Living is risking or, as Lao-tse, the Chinese sage, put it: "Whoever undertakes to hew wood for the master carpenter rarely escapes injuring his own hands." In the past, the broken or fractured bones of early prehistoric man and his forerunners have been attributed, by writers living in an age of violence, to acts of internecine strife. But contemporary anthropologists have shown that such interpretations are unfounded, for predatory animals and the weight of overlying rocks are more than sufficient to explain such breakages. Arrows, first loosed ten millennia before the birth of Christ, remain lodged in rib cages and backbones. Bones broke in forest and furrow as man pursued the chase and tilled the earth. Hunters of the Australian outback and the Pacific Northwest broke arms, legs, ankles and feet scrambling after game. In the rugged terrain of northern Europe, where breaking and cropping ground was hazardous, broken legs were far more common than in the gentler lands of the Nile River valley. The rigorous urgencies of staying alive left scant mercy to the lame.

Ancient Orthopedics

Skeletal remains are lone witnesses to man's first efforts at healing. Diseases like poliomyelitis and arthritis confounded him, but his short life spared him such lingering disabilities. Simple fractures most often healed, though the bones were seldom well aligned. The origins of the splint, like those of the wheel, are cloaked in shrouds of time. Splints of hide and clay, along with the limbs they bound, signify an early intuition that severe breaks must be kept supported and immobile. But, unable to overcome muscle spasms, early man could not align displaced bones well enough to make full use of splints. Bad breaks, though mended, often left wretched deformities. Crude knives, saws and bits are dull reminders of the sharp pains of early surgery. Healed stumps tell modern man of agonizing but successful amputations.

14

*Medieval physicians, painted by
Hieronymus Bosch, cut the skull to
excise "the stone of madness."
Trephination was an international
technique. The Peruvian skull,
bottom, shows signs of healing.*

Among primitive operations none inspires more awe or poses more puzzles than trephination, opening holes in the skull. Although found in widely scattered places, most trephined skulls come from western Europe, especially France — some 10,000 years before Christ — and South America, particularly Bolivia and Peru — several centuries before Columbus discovered the New World. Near several of the skulls lay sharpened stones, generally flint or obsidian, used to make the incision. Techniques varied. Some ancient surgeons drilled the skull by turning a blade on its axis and scoring the bone until a disk could be pried up like a manhole cover. Others cut two pairs of parallel lines at right angles, as though readying to play tick-tack-toe, then levered the square from the center. Still others scraped the skull until a hole opened.

Scholars have long disputed the purpose of trephination. Some consider it solely a ritual procedure, arguing that opening the skull was intended to free demons possessing the patient. Rondelles, scraps of trephined skulls which served as amulets, were sometimes cut from the skull after death, when the operation could not possibly have had a medical purpose. Recently, however, more and more scholars have come to believe trephination was performed to treat injuries, and perhaps disease. Many trephined skulls show signs of fractures, particularly those in Peru where the sling and mace were common weapons. Whether the operation was also undertaken to ease epilepsy remains unknown. Some skulls suggest it may have been performed to relieve severe headaches. The efficacy of the treatment cannot be measured, but it was given with confidence. Most trephined skulls show signs of healing — as many as three-fourths of the Peruvian specimens — and many are riddled with the holes of repeated operations. One Peruvian skull has seven holes, all well healed.

Trephination was part of European medical lore and practice until the Middle Ages, but, from the Stone Age onward, was performed ever less frequently and successfully. For reasons unknown, as man's interest in medicine grew, he lost his inherited skill for performing one of his first surgical operations.

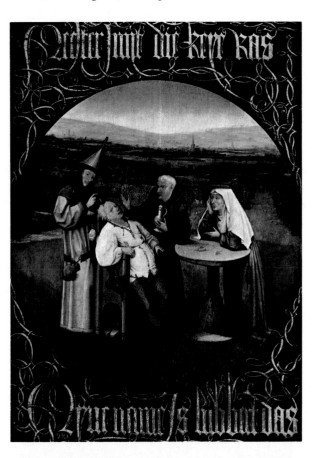

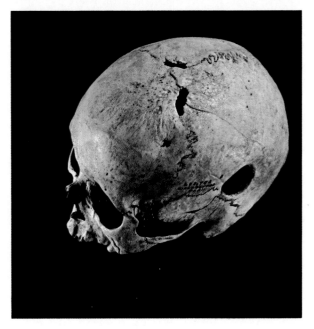

15

Along the valley of the Nile, cradle of Egyptian civilization, the murmurs of bones were first joined by the written word to herald the beginnings of formal medicine. Papyri, prepared between 1900 and 1550 B.C., are the first known medical writings. All are thought to bear the teachings of Imhotep, celebrated physician of the thirtieth century B.C. The Egyptians traced disease to occult agents and wove benevolent magic to cure it. But wounds and injuries suffered in war, farming and building required no magical explanation. Imhotep learned his medicine from experience, by treating the injuries of workmen and studying the craft of embalmers.

One papyrus contains procedures for the treatment of forty-eight different wounds and injuries, most of them affecting bones. Procedures for setting broken bones and aligning dislocated joints, both of which made use of splints, followed reasoned approaches. From ancient Egypt also comes the earliest evidence of the crutch, shown on a tomb carved in 2830 B.C., and the spa, a popular therapy for arthritic pain. Although priests dominated Egyptian medicine, laymen practiced widely in populous centers, sharing the respect shown the profession. Egyptian physicians not only served their patients well, but left a rich legacy to the Greeks who laid the groundwork of medical science.

Man the Measure

Of all the glories of Greece none was greater than man, the centerpiece of classical culture. Man, to the Greeks, was the measure of all things. The Greeks reveled in the human body, crafting it in art, praising it with poetry, testing it in sport and healing it with medicine. Despite their fondness for contemplation and knowledge, nothing so moved the Greeks as the heroic ideal of winning honor through action. Reflected in many aspects of life, this ideal was most tellingly expressed in the Greek taste for war and games, classical stages for physical drama.

Delighting in their bodies, the Greeks prized sound health. A paean, written by Ariphron around 400 B.C., hailed "Health, best of the blessed ones to men," without which "no man is happy." In ancient Greece, the cult of the body

Composed about 1650 B.C., the Edwin Smith Papyrus, named after its discoverer, is among the earliest known medical writings. Its pages tell how to treat forty-eight wounds and injuries to bones and joints. Like kindred papyri, it is thought to express the wisdom of Imhotep, a gifted physician who learned his art healing those hurt building pyramids he designed.

and the desire for health joined with a rational and enterprising spirit of inquiry to spark major advances in medicine.

Greek medicine, especially surgery, began on the battlefield. In the *Iliad,* Homer tells of 147 wounds, often in clinical language, as when the son of Tydeus, hefting a huge stone

> . . . threw, and caught Aeneas in the hip,
> in the place where the hip-bone
> turns inside the thigh,
> the place men call the cup-socket.
> It smashed the cup-socket
> and broke the tendons both sides of it. . . .

Three of four blows and slashes delivered in the epic poem were fatal. Few warriors were spared. "Over these," wrote Homer, "healers skilled in medicine are working to cure their wounds." Greek medicine was an independent art, practiced by devoted and able men who earned their living by it.

This tradition of secular medicine survived the challenge of mystical healing to flower in the writings of Hippocrates. Brought west from the Orient, mystical medicine for a time took hold in Greece in the worship of Aesculapius, son of Apollo and pupil of the centaur Chiron — the founders of healing arts. Temples dedicated to Aesculapius, were the original hospitals. Perhaps the greatest of these establishments lay on the small Aegean island of Cos. Here, four centuries before Christ, Hippocrates worked. Apart from his legendary skill as a physician, little is known of Hippocrates, to whom the first scientific medical writings are ascribed. Written sporadically between the fourth century B.C. and the first century A.D., the works of Hippocrates and his followers leavened the teachings of the Egyptians with critical reason and direct observation. The principles, if not the methods, laid down in the writings inspire medical progress even today.

17

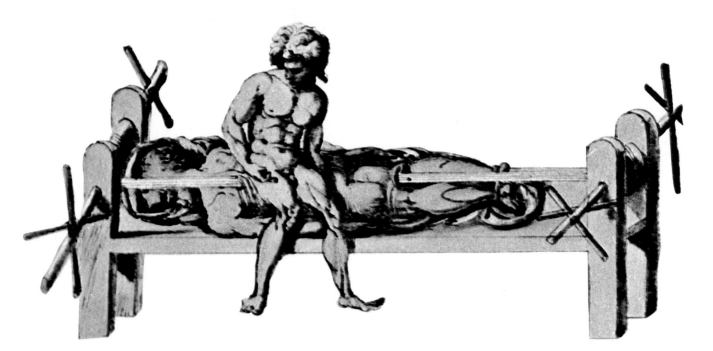

Mending the skeleton calls for strength as well as skill. Equipped with capstans, the Hippocratic bench, or scamnum, above, lent strength to the physician's arm, enabling him to put broken bones and twisted joints back into place. In this drawing from a Byzantine edition of Galen's manuscript, a physician at the bench straightens a patient's backbone. Another Hippocratic procedure, opposite, used the weight of the patient's body, hung upside down on an upright frame, to create the forces needed to realign his spine.

Among the most impressive books in the collection are those dealing with broken bones and dislocated joints. Fractures were treated with bandages and splints. Sometimes clay or starch was used to strengthen the bandage, effectively making it into a cast. Setting a broken leg required great effort to stretch the limb and keep the pieces of the broken bones from overriding one another. Two strong helpers were engaged or an elaborate apparatus, even a light winch, was used to exert the necessary traction.

Several techniques for aligning a dislocated shoulder were offered. By pressing his fist into his armpit beneath the injured shoulder and drawing his elbow to his chest, the patient could sometimes heal himself. The physician could choose any of several methods. He might draw the patient's forearm to his spine then press upwards on his elbow, levering the humerus back into its socket. As with fractures, dislocations of hips and legs required greater force.

The Hippocratic bench, or scamnum, was fashioned of wood and iron. It was designed to exert mechanically the stretching and levering forces the physician required. Fitted with crankshafts and levers and laced with straps and cords, the bench held the patient in position and applied the forces needed to treat him. The bench was no simple device and its precise workings have so far eluded scholars. Medieval Europeans were likewise bewildered by the contraption, though they still found work for it. What had been hewn for healing was now used to cripple and maim — the scamnum became the rack.

As shadow fell over Greek civilization, a false dawn of scientific progress rose over Rome. With a scalpel in one hand and the Hippocratic writings in the other, Galen, during the second century A.D., added experimental dimension to Greek medical literature. His dissections, though performed primarily on animals, furthered the knowledge of anatomy. Chancing upon the skeleton of a robber, picked clean by buzzards, Galen examined it thoroughly. He urged others who would study bones to likewise handle them, heft them, observe their shape and feel their texture. Apart from observing and experimenting, Galen breathed fresh life into the classics of Greek medicine, including their study of the skeleton, by probing inside the body. But with the collapse of Rome, the body of medical knowledge turned, like a skeleton, to something lasting but lifeless, to be venerated rather than nurtured.

Rome, weakened by epidemic and shaken by natural disaster, was overcome by decay from within and assault from without. St. Augustine saw the catastrophes befalling Rome as sure signs of divine wrath visited on those who enjoyed worldly things but flouted spiritual values, who indulged the body but neglected the soul. In its decline and fall, the Roman Empire turned to Christianity for solace.

Medieval Christendom strayed from the ways of classical medical science. Learning languished. Men of the Middle Ages preferred to worship the bones of martyrs rather than study the skeleton of man. While Christianity took hold in Europe, Islam spread across North Africa and the Middle East. Arab ascendancy dimmed the light of classical learning where it glowed brightest, at the famous school of medicine at Alexandria. The Arabs shepherded the libraries and carefully preserved classical writings. But cultural exchange foundered on the bitter rivalry between cross and crescent. Christianity and Islam waged holy war from the seventh century on, leaving Europe bereft of much of its classical heritage. And, like Christianity, Islam fostered mystical attitudes at odds with scientific inquiry.

Although the spirit of classical learning waned, medieval scholars in Europe clung to the letter of what writings they possessed. One of the fore-

20

most medical writers of the early Middle Ages, Paul of Aegina, writing in the seventh century, clearly expressed this reverence for the ancients who "said all that could be said on the subject." When explaining ambiguities or filling gaps in ancient texts, however, medieval writers frequently offered original suggestions. No exception, Paul of Aegina, though a humble disciple of the ancients, garnished their teachings in several ways. Bolder than Hippocrates in treating spinal disorders, he suggested relieving paralyzing pressure on the spinal cord by surgery. Likewise, he introduced new techniques for mending shattered kneecaps. His notes on fractures reveal his experience. Paul recommended massaging, even cutting the callus, fresh bone tissue that formed around a poorly set fracture. The treatises of Paul of Aegina, particularly those on surgery, stood alongside Hippocratic and Galenic writings as the staples of medicine for centuries.

Like other orphaned children of classical culture, medicine found a home in the monasteries where friars and monks tended the sick and lame. Inspired by St. Benedict, monastic medicine

Their mischief done, demons lurk and wheel in the background as St. Benedict revives a stricken monk in this fourteenth-century Florentine fresco. Although urged by their founder to care for the sick, the Benedictines were forbidden to study medicine for St. Benedict believed God alone could cure man's ills.

21

flourished at Monte Cassino in Italy, Tours in France, St. Gall in Germany and Oxford and Cambridge in England. But until its decline in the fourteenth century, the school at Salerno in Sicily was unmatched as a center of medical inquiry.

Pride of Medieval Medicine

Although begun by the Benedictines, Salerno bore a secular stamp. Lying at the crossroads of the Mediterranean, the school attracted gifted pupils and teachers who, during the eleventh and twelfth centuries, began the return to classical tradition which led to the Renaissance. Anatomy, studied by dissecting animals, dominated instruction at Salerno. Since Rome discouraged dissection and Islam forbade it, the revival at Salerno marked a breakthrough in anatomical inquiry. Dissections of human bodies, usually those of executed criminals, began to be performed. Holy Roman Emperor Frederick II made dissections mandatory at the Italian schools in 1240, much to the dismay of the Pope. The school became a channel through which Greek and Roman texts, long hostage of Islam, flowed back into Europe.

The rise of Salerno was but one thread in the rich tapestry of the Renaissance. Taken to new worlds by the crusades, Christendom found what the Arabs called "the science of the Greeks." From Sicily, Byzantium and Spain the writings trickled into Europe, arriving just as Christendom was in the throes of crisis. Pope Gregory VII, seeking to assert the authority of the papacy against the power of the emperor, opened the struggle between church and state in the eleventh century. In waging the contest, the Catholic Church retreated into the redoubt of religious orthodoxy, leaving new learning to laymen. In the twelfth century monks were forbidden to practice medicine, a measure which spelled the end of monastic medicine. But the great European universities of the Renaissance emerged to take up the cultural slack. At Paris and Orléans, Oxford and Cambridge, Bologna and Padua, the ancients, orphaned once more, were taken in.

From Salerno, the center of medical studies moved northward to Bologna, where Guglielmo de Saliceto pursued anatomy and surgery in the

Drawn from the four corners of the Mediterranean, doctors of Salerno, shown in this woodcut dining and working in the same hall, were the most gifted and daring of the Middle Ages. The regimen of Salerno, enriched by the flow of classical writings and enlivened by a spirit of rational inquiry, grew into a tradition respected across Europe.

22

tradition of Hippocrates and Galen. His masterpiece, the *Cyrurgia,* presents little original work but marks a thirteenth-century return to classical methods. Saliceto was the first European to link crepitus, the sound made by the ends of broken bones rubbing together, to a fracture. Knowledge of the neck enabled Saliceto to realign dislocated vertebrae without resort to mechanical devices. Whether he gained his understanding of anatomy from dissections is unclear. A few dissections were probably performed. One Bolognese physician, Mondino de Luzzi, remarked that bones were best studied after boiling them, adding for safety's sake that "to do so would be sinful."

By the fourteenth century the Franciscans and Dominicans were leading a counterattack against religious heresy and intellectual innovation. One of Saliceto's prize pupils, Lanfranchi di Milano, better known as Lanfranc, ran afoul of authority and fled to France. With Lanfranc, the center of European surgery moved to France where it was to stay for several centuries.

Son of peasants and physician to popes, France's Guy de Chauliac bridged the gap between medieval and Renaissance medicine. After a secular education, including a period of instruction at Bologna, he entered the Church, ultimately to serve three popes at Avignon between 1342 and 1370. His major work, the *Chirurgia Magna,* is a vast compendium of medical lore. Like all schooled in the Bolognese manner, de Chauliac stressed anatomy, insisting that surgeons ignorant of it carved the body like the blind carved wood. Although he witnessed and performed numerous dissections, de Chauliac was a less than accomplished anatomist. Despite his own maxim, however, he was a skilled surgeon.

At Bologna he learned to favor simple, light splints fashioned of willow, leather or horn over other contrivances as ineffective as they were cumbersome. He pioneered the use of weights and pulleys for holding broken legs in constant traction, a treatment not commonly given until four centuries later. Nor did he shrink from rebreaking and resetting fractures which had failed to mend properly.

Apart from his contributions to surgical practice, de Chauliac did much to enhance the place of medicine in the eyes of a suspicious Catholic Church. In defiance of clerical authority, dissections were performed every two years, beginning in 1340, in the anatomy class at Montpellier, where de Chauliac taught. But Pope Clement VI, spared the ravages of the Black Death in 1348 by de Chauliac's treatment, encouraged dissection "in order that the origins of disease might be known." Although dissection would fall under official shadow again and again, Clement's initiative reflected man's changing conception of himself and his body.

No one captured this change more surely or rendered it more finely than the artists of the Renaissance. By the fifteenth century, they boldly pictured the human body's vitality and beauty, frequently unfettered by clothing. Artistic inspiration enhanced anatomical knowledge as never before, and in no one more keenly than Leonardo da Vinci.

More versatile than revered, barber-surgeons often suffered the slings of satirists. Pictured as beasts, they cavort about this engraving letting blood, pulling teeth, cutting hair and dressing wounds while an apprehensive patient is ushered into their midst.

Leonardo not only studied the skeleton but was the first to picture it accurately in its entirety. Said to have dissected thirty men and women, Leonardo relied almost solely on his senses, paying scant heed to written sources. Perhaps early experience in engineering led him to perceive the skeleton as a set of rigid levers "of inflexible hardness adapted for resistance and without feeling." From his profound understanding of muscles, Leonardo grasped the operation of these levers. In his drawings of the skeleton, bones are joined by cords, as they were in his models, to illustrate lines of muscular tension. First and foremost an artist, Leonardo knew that the deeper the understanding of function, the truer the portrayal of form.

Surgery at War

If medicine owes something to the beauties of art, it owes even more to the horrors of war. The introduction of artillery, first fired at Crécy in 1346 during the Hundred Years War, fundamentally changed the art of war. During the Middle Ages knights, bowmen and pikemen closed in hand-to-hand fighting which left only the quick and the dead. With artillery and musketry, larger armies, engaging at a distance, inflicted more

26

wounds but fewer fatalities. Between the sixteenth and eighteenth centuries, as armies warred across Europe, field surgeons followed in their trains. Coping with myriad injuries, they overcame their poor schooling in medical theory and often enriched theory with experience.

Barber-surgeons were another source of practical surgical knowledge. Despised by physicians and scholars — and often feared by patients — barber-surgeons learned their trade by practicing as apprentices and journeymen, like other master craftsmen. Quality of training and dedication to craft varied. Samuel Pepys, the playful London diarist, attended a dissection, or "anatomy," held by the United Company of Barber-Surgeons, where everyone wined and dined, ignoring their guest of honor. Yet from the ranks of the French barber-surgeons came a pioneer who forever changed the practice of surgery.

Medicine's Best Barber

Ambroise Paré, apprenticed as a boy of sixteen in 1526, healed countless soldiers and served four kings in raising his craft to an art. Paré's major work, *Dix Livres de la Chirurgie,* reflected an understanding of anatomy drawn heavily from the works of others but contained many original surgical insights. Paré's single most important innovation was the use of ligatures, or tourniquets, in the performance of amputations. By tying a ligature just above the line of incision, the muscles and skin were held up, later to be drawn over the severed bones to "serve them instead of bolsters or pillows when they are healed up." The ligature also stanched blood and dulled pain. By replacing cauterization, Paré's procedure eased the immediate agonies of amputation. And it left a stump much less painful and far better suited to an artificial limb. Paré himself designed artificial limbs, including a leg with a hinged knee joint.

Devised in classical times, artificial limbs enjoyed a vogue during the Renaissance. Paradoxically, just when the knight in armor was going the way of Don Quixote, tilting at windmills rather than warriors, the craft of the armorer reached its peak. For these artisans it was a short step from fashioning exquisite protective armor to making ingenious mechanical limbs when the

Ambroise Paré

A Soldier's Surgeon

"Dare you teach me surgery?" French physician Ambroise Paré raged at a professor who questioned his abilities. "Surgery is learned by hand and eye," he insisted, before closing with a withering rebuke. "You, my little master, know nothing apart from how to chatter in the chair."

He launched his scornful broadside in 1585, when he was seventy-five years old. More than fifty years a surgeon, he was often snubbed and censured by the French medical establishment for championing new techniques and challenging archaic authorities like Hippocrates and Galen. Neither ruffled nor daunted by criticism, he relied on experience to become the master surgeon of the sixteenth century.

When Paré first apprenticed as a barber-surgeon in the French countryside, his craft was the despised stepchild of medicine. Learned physicians disliked shedding blood, except to purge the body of unbalanced "humors." They left surgery to barbers and tinkers, even executioners. After a brief spell as resident surgeon at the Hôtel-Dieu, then Paris's only public hospital, Paré joined the army, serving in the religious wars which scarred the century. He

honed his surgical skills on many campaigns, matching his wits against "wounds made by gunshot and other fiery Engines [made] for the speedy and cruel slaughter of men."

In 1536, at the siege of Turin in northern Italy, Paré stumbled on his first major medical discovery. At the time, bullet wounds were treated with scalding oil, but after a long day of surgery, Paré had run out. He dressed wounds with a salve of egg yolks, turpentine and rose oil. He slept fitfully, fearing to find his patients dead come morning. Instead he was surprised to discover that his

patients' wounds had healed cleanly and with little pain. He resolved "never again to burn thus so cruelly the poor wounded." On another battlefield, Paré first used silken ligatures, or tourniquets, when amputating limbs, instead of cauterizing wounds with hot oil. If necessity be the mother of invention, Paré proved himself an ingenious midwife.

He ministered to injuries of misfortune with the same skill as wounds of malice. Sifting tradition with experience, he refined procedures for setting fractures and straightening dislocations. Paré's sympathy was aroused by spinal deformities, particularly in children. Singling out "the unhandsome and indecent situation of their bodies when they are young and tender, either carrying, sitting or standing . . . sewing, writing or any such like thing," he recognized the dangers poor posture posed to the skeleton long before such a diagnosis became common.

Trusting experience, and using dogma and theory only to confirm it, Paré stood near the threshold of modern medical science. He defied critics and suspected authorities, holding that medicine was the handmaiden of nature. "I dressed him," Paré said humbly, "God healed him."

The two finest physicians of their time, Ambroise Paré and Andreas Vesalius, confer vainly at the death-bed of Henry II of France in 1559, below. Three of Henry's sons ruled after him and Paré served them all.

Enamored of instruments and devices, Paré designed both. His sketch, bottom, is of an artificial leg. Despite ingenious design and fine workmanship, artificial limbs, like armor, were heavy and cumbersome.

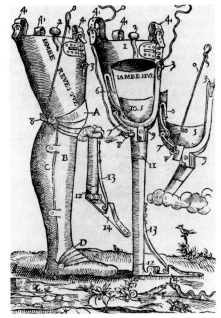

armor failed. Artificial limbs were forged of metal to resemble as nearly as possible the lost original, even down to the fingertips. Goetz von Berlichingen was literally an iron-fisted champion, able, by the poet Goethe's account, to wield a deadly sword. But, despite the ingenuity of design and care in construction such devices were probably more cumbersome than effective.

Paré, generally following traditional methods of diagnosis and treatment of fractures and dislocations, often supplemented them with the use of one device or another. Keenly interested in scoliosis, lateral curvature of the spine, he recommended steel corsets, the forerunner of later spinal braces and plaster jackets. He also devised walking splints for those crippled by ailing hips and special shoes for those hobbled by clubfoot. A wide range of surgical instruments, among them a crow's beak used as an arterial clamp, reflected Paré's resourcefulness. With wit, talent

and wisdom he brought respect to a craft loathed and feared when he took it up.

Surgery became legitimate just as Europe became enthralled by science. During the seventeenth century the Scientific Revolution, with astronomy in the vanguard, vindicated the principles and methods of empirical science. While pioneers like Johannes Kepler, Galileo Galilei and Isaac Newton confirmed Copernicus's heavenly reconnaissance, Vesalius mapped the human body in brilliant anatomical explorations.

Although anatomy and pathology led the way, surgeons began to heal the rift dividing them from the rest of the medical establishment. They increasingly turned to easing the distress of those marred and crippled by skeletal deformities. During the Middle Ages, physical deformities invited scorn, for they were considered signs of the scourge of God. Medieval artists drew the minions of Satan — demons and imps, warlocks and witches — with hunched backs, crooked limbs and twisted bodies. Hunchbacks and dwarfs became so popular as royal entertainers that parents deliberately maimed their children in the hope of finding them a jester's job at court. Sources of curiosity well into modern times, deformities began to attract greater sympathy in the seven-

teenth century. Rodriguez de Silva y Velázquez painted dwarfs at the Spanish court with sensitivity and Neapolitan José Ribera captured the torment of polio in his painting of a beggar boy. While some comforted, others healed.

Many sought to correct misshapen backs and limbs. In Germany, Wilhelm Fabrig presented the first pictorial descriptions of curvature of the spine from anatomical sections of a child's vertebrae, showing the deformity's full complexity. He also treated clubfoot with a boot of his own design, an adjustable device that corrected the deformity gradually. Fabrig, along with Paré and others, also attempted to overcome clubfoot with braces. But the full effectiveness of these treatments awaited a greater understanding of the origins and progress of the deformities themselves.

For this, surgeons depended upon their colleagues studying more theoretical aspects of medicine. English anatomist William Harvey studied the growth and development of the embryo, opening the way for a closer understanding of congenital abnormalities. Meanwhile, Italian anatomist Marc Aurel Severinus firmly established the link between spinal deformities and tuberculosis. Thomas Sydenham, "the English Hippocrates," described how rheumatism, gout, scurvy and dysentery damaged bones and hindered joints. In Holland, Henrick van Deventer, an accomplished surgeon and obstetrician, found many deliveries hampered by deformed pelvises, a considerable number of them caused by scoliosis, or lateral curvature of the spine. Turning to skeletal deformities, van Deventer recorded precise descriptions of the common abnormalities of the pelvis and spine.

The most impressive of these studies, on rickets, was the work of English physician Francis Glisson. Although he knew nothing of its causes, he correctly interpreted its deformities as arising from the unbalanced growth of bone. He distinguished rickets from infantile scurvy, a refinement overlooked by his successors until the nineteenth century. Like his contemporaries, he attempted to correct the disease's deformities by using braces, splints and shoes to straighten twisted joints and bent limbs. Glisson supplemented the application of these various devices with a carefully prescribed regimen of exercise and massage to strengthen weakened muscles. Despite steady progress in anatomical knowledge, surgical technique and therapeutic methods, the treatment of skeletal disorders waited upon more thorough understanding of the nature and character of bone itself.

Charting Bone's Canals

Toward the close of the seventeenth century, bone began to surrender its secrets to the microscope. Englishman Clopton Havers devoted himself singlemindedly to examining bones and joints. He was the first to explore the internal structure of bone. He discovered channels, today known as Haversian canals, running along the shafts of the long bones of the arms and legs. Havers could not see, nor did he suspect, that the channels carried capillaries which nourished bone. He also suggested that the periosteum, the membrane surrounding bones, was sensitive to biological processes occurring within the bone, a conclusion confirmed two-and-a-half centuries later. Havers described the properties of cartilage, the coarse tissue cushioning bones, and the synovia, a clear fluid that lubricates the joints. Taking his observations as far as he could without more

At the court of Philip IV of Spain, Velázquez found a dignity in the less fortunate — jesters, beggars, idiots and dwarfs — that other artists had reserved for their privileged patrons. In The Maids of Honor, *painted in 1656, he cast Maria Barbola, a dwarf, bearing herself as proudly as any queen. Velázquez's sensitivity signaled growing understanding and respect for people scorned in earlier centuries as freaks and demons.*

31

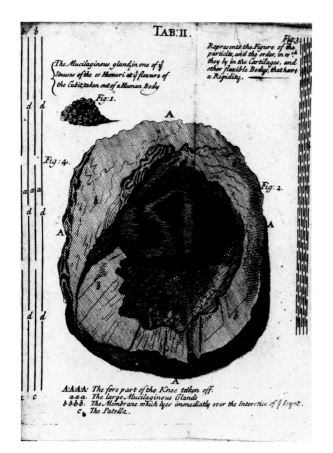

English anatomist Clopton Havers pioneered scientific inquiry into the structure of bones and joints in his Osteologia Nova of 1691. This illustration from Havers's text highlights the kneecap and cartilage.

detailed knowledge of the cell, still over a century away, Havers laid the groundwork for the study of the physiology and pathology of bones.

During the next century the study and treatment of the skeleton advanced as a clearly defined specialty within medical science. Professor of Medicine at Paris Nicholas André gave this branch of medicine the name orthopaedia, from the Greek words *orthos,* which means straight, free from deformity, and *paidios,* a child.

André spent his career seeking to correct skeletal deformities. Convinced the symmetrical growth of the skeleton depended on muscular balance, he traced deformities to poor posture during the growing years. "It is well worthwhile to remark that Crookedness of the spine does not always proceed from a fault in the spine itself," he wrote, "but is sometimes owing to Muscles of the forepart of the Body being too short, whereby the spine is rendered crooked, just in the same manner as a Bow is made more crooked by tying its Cord tighter." Both Hippocrates and Paré had reached similar conclusions. But neither devised corrective measures to rival those of André.

He presented ways to prevent and cure curvature of the spine. Generations of schoolchildren, sternly admonished to sit up straight in hard-bottomed, stiff-backed chairs, might well spare the schoolmarm and curse André. He even specified the relative heights of chairs and tables to ensure proper posture at mealtimes. To restore the balance of muscles and straighten the spine, he recommended a regimen of exercise and rest and the wearing of corsets and braces. Others followed his lead, mixing posture training, physical therapy and mechanical devices in different measures. At Orbe, Switzerland, in 1790, physician Jean André Venel established the first institute entirely devoted to the diagnosis and treatment of skeletal deformities.

A similar desire to ease disabilities and prevent crippling deformities prompted Peter Camper, an Englishman well versed in anthropology and anatomy, to write *Dissertation on the Best Form of Shoes* in 1781. He introduced to the medical profession the problems arising from the relationship between the mechanics of the feet and the nature of shoes. These problems have consumed

In his Orthopaedia *of 1743,*
French physician Nicholas André
suggested that a lady might correct
her elevated shoulder by carrying a
weight and a gentleman prop up his
deformed shoulder with a ladder.

much of the time of orthopedists ever since, but little of significance has been added to the original approaches presented by Camper.

In the eighteenth century, the treatment of skeletal deformities benefited greatly from mounting knowledge of the growth and structure of bone. At Guy's Hospital in London in 1763, surgeon John Belchier successfully measured new bone growth by adding dye to the diet of animals. During the mid-1700s, Albrecht von Haller, a Swiss physiologist, and Henri-Louis Duhamel, a French surgeon, advanced rival explanations of the growth of bone. Haller reckoned the blood stream carried bone to the periosteum where it gathered and grew. Duhamel argued that bone grew from the periosteum itself, much like the wood of a tree grows from its bark. From these competing theories sprang two parallel lines of research reaching, without resolution, well into the twentieth century.

In England, the Hunter brothers, William and John, studied every aspect of skeletal anatomy, physiology and surgery. William's work on the nutrition of cartilage, for all its originality, was overshadowed by the research of his younger brother. Among the findings which flowed from John Hunter's laboratory was the novel conclusion that bone was living tissue that grew, mended and changed under life's stresses and strains. "The bones, in their causes of disease, in many of their diseases, in the termination of these, and in their restoration," he noted, "are similar to soft parts. . . . Nor can they require a different treatment, as regards their vital power." An able surgeon, Hunter was dubbed by one biographer "a prince to the thinking surgeon and only a babbler to the merely practical ones."

As knowledge of the physiology of bone mounted, surgeons operated with growing confidence, venturing new procedures and extending tried ones. English surgeon Charles White removed four inches of diseased bone from a humerus which, after four months, grew enough to enable the patient to use his arm again. The port of Liverpool teemed with sailors whose limbs, maimed on distant voyages, were often spared by the skilled hands of Henry Park. Loath to amputate, Park developed a number of operations for

33

Wilhelm Roentgen

Turning the Key to the Skeleton

One fall day in 1895, Wilhelm Roentgen was in his laboratory at Würzburg, Germany, experimenting with vacuum tubes. Charged with electricity, the tubes gleamed and flickered, casting iridescent hues around the dim room. His work interrupted, Roentgen put a tube, still aglow, on a book. Between the pages, marking his place, lay a large antique key. The book rested atop one of the photographic plates he kept close at hand. Deciding to spend the afternoon taking pictures along the Main River, Roentgen returned to his laboratory and gathered up his plates, among them the one under the book. Later, when he developed his exposures, he was puzzled to find clearly etched on one plate the shadow of the key.

Putting other work aside, Roentgen conducted a series of experiments to explain the mysterious appearance of the key. He began by darkening his laboratory and covering a vacuum tube with a black cardboard box, to trap any light generated by the tube. Ten feet away, he placed a screen coated with a fluorescent chemical compound. When Roentgen turned on the current, the screen shimmered a ghostly green, dancing to the rhythm of the charges passing through the tube. Some mysterious form of energy had fled the tube, eluded the eye and reached across the room to light the screen.

In weeks of earnest pursuit, Roentgen captured and questioned the fugitive source of energy. Guessing that if it passed through air it might also breach other things, he placed several obstacles — paper, metal, rubber and wood — between the tube and the screen. It penetrated them all, except a sheet of lead. Intrigued by the stopping power of lead, Roentgen held a small piece in his hand. Looking across the room, he was stunned to see, thrown on the screen, shadowy images of the bones of his thumb and finger. Finally Roentgen aimed the rays at his wife's hand. On the developed plates, Frau Roentgen's bones, and a ring, shone a ghostly white against the dark background of her flesh.

Soon after New Year's Day, 1896, Roentgen announced the discovery of X-rays. At the request of one of his colleagues, an anatomist, he agreed to explore their medical uses. Roentgen's rays at once sparked a furor around the world. Meeting widespread fears for ladies' modesty, a London firm advertised lingerie impervious to the leer of X-rays. Across the Atlantic, the New Jersey legislature considered banning the manufacture of X-ray opera glasses. Prohibitionists greeted the discovery more soberly and prophetically, hoping it would reveal the damage wreaked by tobacco and alcohol and so discourage their use.

Awarded the first Nobel Prize in physics in 1901, Roentgen was given a more moving tribute when, during World War I, surgeons on both sides toasted his birthday. Early in his career, Roentgen had remarked that the scientific experiment "is the most powerful and reliable lever enabling us to extract secrets from nature." His own experiments, following on a chance picture of a skeleton key, revealed a secret which has disclosed secrets ever since.

removing shattered and diseased joints while saving the limb itself. In contrast, amputation was routinely performed by army field surgeons until Park's influence spread to save limbs his hands could not reach.

Surgeons also set fractures and realigned dislocations with increasing success. Percival Pott of St. Bartholomew's Hospital in London gave his name to the common break above the ankle. Pott's accurate mapping of skeletal anatomy enabled him to identify the precise site of fractures and dislocations so that he could set and align them with greater chances of full recovery.

Treatment of diseases, however, did not keep pace with improvements in surgical techniques. Many bone diseases had long been recognized, but measures to prevent and treatments to cure them developed slowly so long as the causes eluded researchers.

Beneath the Skin

Two discoveries transformed prevailing concepts of disease in the nineteenth century. Rudolph Virchow, a German pathologist, formulated his doctrines of cellular pathology. He taught that the source of disease should always be sought inside the cell and that symptoms of disease are but the reactions of cells to its cause. Although he exaggerated his case, as many pioneers do, Virchow placed pathology on a sound footing. His French contemporary Louis Pasteur discovered that infection was the body's response to invasion by tiny living organisms — bacteria.

But many diseases, particularly of the bones, remained unknown, even unseen. Then, in 1895, German physicist Wilhelm Roentgen, quite by accident, discovered a new form of energy which he called X-rays. For the first time, living bones were laid bare. Physicians could see breaks, wear, tumors and lesions. They could trace the subtle changes occurring in the living skeletons of their patients. Through this unique window into the body, bones became like flesh, visible, obvious, accessible to the physician's practiced sight. By the beginning of the twentieth century, with Roentgen's radioactive vision, doctors could unmask diseases whose only trail, for millennia, had been suffering.

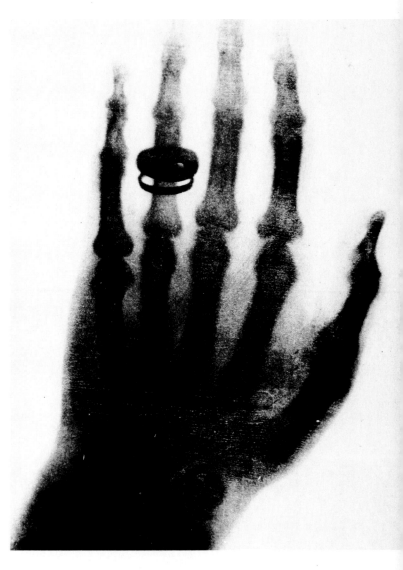

Chapter 2

The Flexible Framework

Moonlight falling upon a Belgian landscape one autumn night in 1536 illuminated the skeleton of a man chained to a gibbet on a country road outside the town of Louvain. Like other crumpled bodies surrounding it, the bones on the post were the remains of an executed criminal. Left on the outskirts of town, they served to remind passing criminals of the risks of a dubious profession.

Amid this desolation a determined figure strode, a dark-eyed young man named Andreas Vesalius. To Vesalius, a dedicated student of anatomy, the abandoned bones were a treasure, an illicit one. Decree forbade the removal of a body from its place of rest, even though it was forsaken. Undaunted, he "took advantage of this unexpected but welcome opportunity." With a few firm yanks, he wrenched all the limbs from the body and smuggled them into town. The head and trunk, still soundly chained to the post, were the objective of another foray. "So great was my desire to possess those bones that in the middle of the night, alone and in the midst of all those corpses," he recalled, "I climbed the stake with considerable effort and did not hesitate to snatch away that which I so desired."

The object of Vesalius's desire was not the skeleton so much as the knowledge it promised. He thought that to understand the anatomy of the human body one must begin with its framework, building knowledge layer by layer, until one had reached a dynamic whole. It was an elegantly simple approach to anatomy that made the classic drawings of the skeleton appearing in his works so lifelike — rendered as though they still had flesh and soul.

Indeed, the skeleton is vital to many forms of life. Without it, movement would be impossible for man. It provides a base for attaching muscles and the leverage to assist in their pulling. The skeleton protects the body's vital organs and pro-

Rigid yet flexible, the skeleton lies at the center of the cascade of motion depicted in Marcel Duchamp's 1912 painting, Nude Descending a Staircase, No. 2. *A complex structure of 206 individual parts, the skeleton is the framework of the body's form and its means of motion.*

A massive bony skull encased the brain of the Dunkleosteus terrelli, *a fish that lived about 350 million years ago. Below, a shark's teeth are remnants of an external skeleton worn by its early ancestors.*

Skeletons, reflecting Nature's diversity, support not only different bodies but vastly different ways of life. The delicate bones of bird and the towering bones of elephant must both fight the force of gravity.

vides life-giving substances — blood cells from red bone marrow and minerals from its bony storehouse. An ever-changing structure, it is also sensitive to its user's needs. The skeleton keeps pace with rapidly growing children and adapts itself to our lifestyles, reinforcing areas where we, from sport, hobby or occupation, add to the forces that already burden it.

The skeleton's most important function is that of support. Like beams that hold up buildings, the skeleton holds the body erect. But the body does not remain rock-still like a building. Where bone meets bone, joints called articulations make the skeleton flexible and give it movement.

Support is perhaps the skeleton's oldest function, one that arose in the waters of ancient times. Scientists trace the first skeletonlike structures to animals that lived some 500 million years ago. Their bodies encased in bony plates, flat jawless fish hovered over the muddy bottoms of Ordovician seas sucking up debris for food. Such external armor, however, limited growth and inhibited movement. Nature, always seeking to better her products, began forming an alternate design. Over a span of about 100 million years, primitive internal skeletons appeared. Backbones developed to give the fish strength and flexibility and bony plates grew to protect their developing brains. The internal skeletons were a success. Fish so equipped were more agile and better able to escape their predators.

Grappling With Gravity

These early skeletons did not shoulder the burden of support alone, but shared it with the seas. The same holds true for the skeletons of modern sea-dwelling creatures. The blue whale, largest animal ever to live, can weigh up to 150 tons. It moves with the grace of a leaf floating on air, for the ocean buoys the giant mammal's weight by meeting it with an equal force to give it external support. If the whale were to wash ashore, it would suffocate within minutes because its skeleton could not keep the whale's weight from crushing its lungs. No skeleton could carry such weight on land and still remain flexible.

Some 150 million years ago giants did walk the earth. Yet the largest dinosaur, the brachiosaurus,

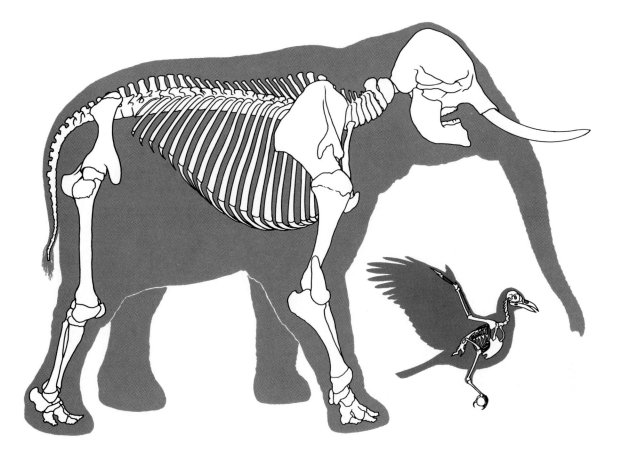

was only one-third the weight of the blue whale. It could walk on land but spent most of its life in water. Its sturdy skeleton is a testament to the battle it fought with gravity. The giant reptile walked slowly and deliberately, scarcely bending its legs. The brachiosaurus was so heavy it could not leap. That was just as well, however, for even a short fall would have seriously injured the heavy beast.

Another mammoth of the sea, the giant squid, has no internal skeleton at all. The largest ever caught measured seventy-two feet in length, but scientists speculate this mollusk may grow even larger. And the powerful great white shark, sometimes growing to thirty-six feet in length, has a skeleton made of cartilage. Its hardest tissues are jagged triangular teeth and the studs on its sandpaperlike skin.

The skeleton became more crucial for animals that lived on land or in the air. Without water, such animals relied solely on the skeleton for support. As demands grew more diverse and complex, so too did support, giving rise to the myriad skeletons in nature today. Like sea animals, both bird and elephant must conquer the force of gravity. The bones of the bird, designed for flight, are as light and streamlined as possible,

giving it the same grace in air that fish have in the sea. The African elephant, largest of the land animals, can reach a weight approaching eight tons. For support, bones in its legs are as sturdy as the trunks of small trees; those in its skull, immense and thick to allow for the attachment of dense muscles that work its cumbersome jaw.

The skeleton of man, too, is tailored to his needs. It allows him neither flight nor awesome stature. The human skeleton is light yet strong. Seemingly simple in structure, it is nonetheless a masterpiece of architecture. Anatomists divide it into two broad categories — the axial and appendicular skeletons. The axial skeleton comprises the bones of the body's central core: the skull, spinal column and thorax, which shelters the heart and lungs. The appendicular skeleton includes all bones in the arms and legs as well as the shoulder and pelvic girdles, which affix the limbs to the axial skeleton.

The adult skeleton totals 206 bones. This number is but an average, however. Some people are born with one extra pair of ribs, some with one pair less. Similarly, the number of bones in the spine can vary. And bones that generally fuse together during life can remain separate. At birth, babies have about 350 individual bones. This

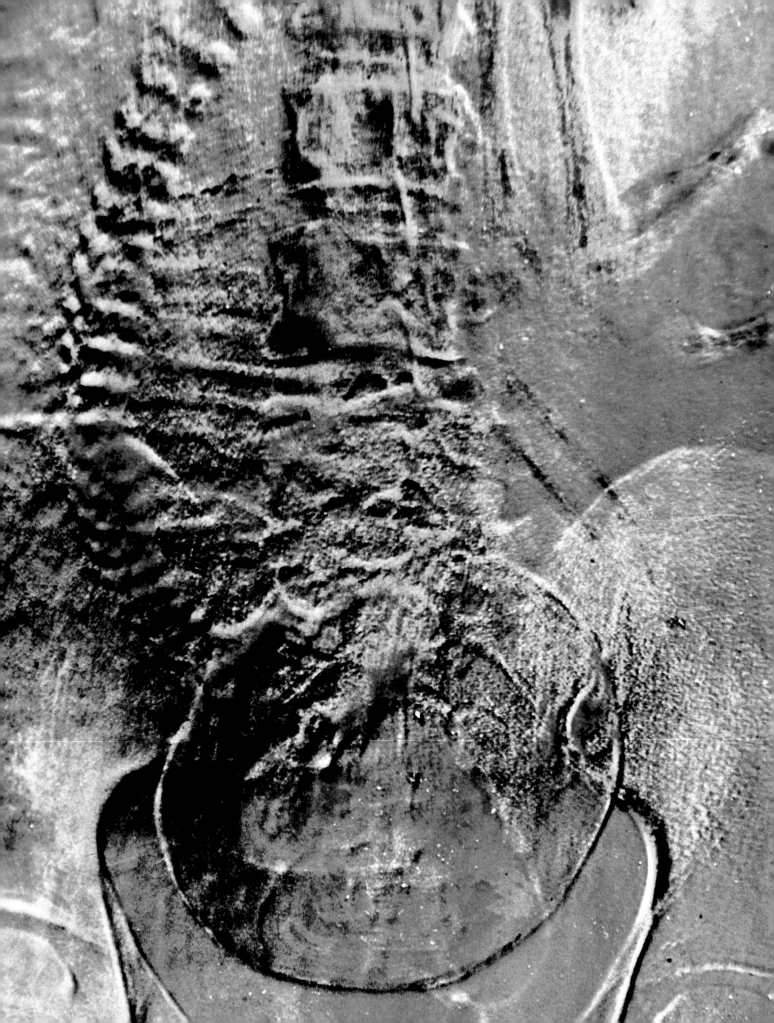

number itself differs depending on which ossification centers — the areas where bone first forms — are counted as individual bones.

In a human embryo only weeks old, specialized cells are already massing. Once positioned, they rapidly begin converting into cartilage. By the time the cartilage is well under construction, bone begins to replace it. Distinct skeletal parts are recognizable in an embryo five weeks old. Even in this early stage when the embryo is about the size of a pea, the skeleton has a prominent spinal column — a single majestic curve that follows the gentle form of the embryo.

A Multifaceted Column

The mature spine is a series of alternating convex and concave curves that support the body and absorb shock. The column is built of thirty-three small bones, the vertebrae. Seven cervical vertebrae in the neck are the smallest and allow the widest range of movement along the spine. Twelve heavier bones, the thoracic vertebrae, lie beneath them, forming the upper back. These bones also hold the ribs in place. Special dish-shaped indentations called facets help anchor the ribs on both sides of the vertebrae. The five lumbar vertebrae that make up the small of the back are the largest bones of the spine and bear most of the body's weight. Beneath these, five smaller bones, individual at birth, fuse at about age twenty-five to form the wedge-shaped sacrum, a single bone fitted between the two hip bones.

Old English folk stories tell that in punishment for the twelfth-century murder of Thomas à Becket, archbishop of Canterbury, the people of Kent were cursed to be born with tails — a sign of kinship with the devil. Every human being does have a tail, or the vestige of one. Tucked beneath the sacrum, the coccyx is a small, tapering bone built of four fused vertebrae. It is the only one without a function.

With the exception of the fused vertebrae in the sacrum and coccyx and two highly specialized vertebrae at the top of the spinal column, all are structurally similar. The main portion of a vertebra, called the body, is a flattened, oval block of bone. Short columns, the pedicles, arise from the back of the body. Small plates called

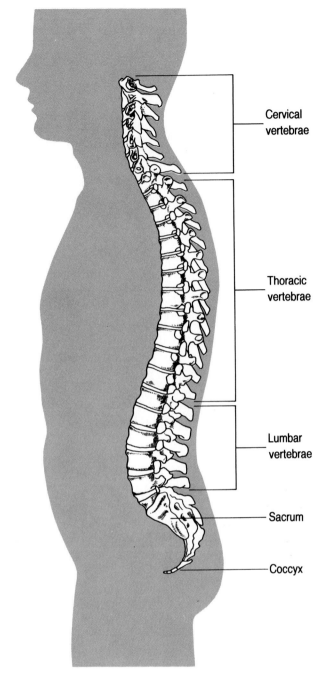

Cervical vertebrae

Thoracic vertebrae

Lumbar vertebrae

Sacrum

Coccyx

The adult spine, engineered for strength and flexibility, curves gracefully, like an archer's bow. The thirty-three vertebrae, so alike in the fetal skeleton, have become specialized, their differences shaped by the requirements of support and movement. Convex curves, thoracic and sacrococcygeal, remain from the original shape of the fetal spine but the two concave curves, cervical and lumbar, are acquired after birth.

A single lumbar vertebra sprouts bony wings called processes that link it to adjacent muscles, below. Severe lateral curvature, scoliosis, disrupts the spine, crowding ribs and restricting chest capacity, right.

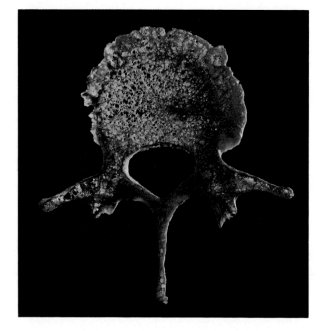

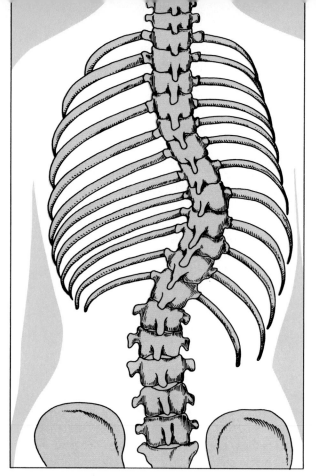

laminae seal the opening between the two pedicles, creating a circular channel through which the spinal cord passes. Three winglike extensions crown the laminae. Spaced at ninety-degree angles from each other, they anchor muscles and give the spine its knobby appearance under the skin. The articular processes, two smooth nubs on top of the vertebra and two beneath, link a vertebra with its neighbors.

Anatomists dubbed the topmost vertebra the atlas, for the Titan in Greek mythology who was condemned to carry the earth and heavens on his shoulders. This bone, supporting the skull, does not have the solid body of bone typical of the other vertebrae. Rather, it forms a bony ring with a large central opening. Rounded projections of the skull's occipital bone — the bone making up much of the skull's rugged base — fit into two large hollows atop the atlas. Tough ligaments bind skull to spine. Beneath the atlas, another vertebra called the axis sends a bony projection through the opening in the atlas. This articulation permits the atlas, holding the skull, to rotate.

All along the spine, strong ligaments, muscles and facets on the vertebrae bind the spine into a single column. Between the vertebrae, there is special padding. Occupying about one-fourth of

an adult's spine, disks absorb shock and prevent the bones from grinding against each other. They allow motion between the vertebrae, bolstering the spine with added flexibility and strength.

The disks also play a role in shaping the spinal column, which rests at the body's vertical center of gravity. The spine crosses the gravity line, weaving back and forth. The resulting curves provide much more stability and strength than a straight column could. To create curves, the disks subtly change shape, with portions narrowing so that the vertebrae do not sit directly on top of each other but rest at a slight angle. The overall effect is a series of alternating concave and convex curves that run gracefully down the spine.

Its distinctive shape is evidence of the spine's adaptation to man's erect stance. Anthropologists have ascribed many advantages to bipedalism, the ability to walk on two feet. Most significantly, they believe bipedalism led to an increase in man's brain size because it freed the hands for tasks more complicated than bearing weight. The history of the spine's adaptation to bipedalism is encapsulated in the experience of every child learning to hold his head upright, to pull himself forward, to crawl and to walk. At each stage, the spinal column responds to new demands.

A kind spirit housed in frightening form, Quasimodo casts untrusting eyes over his deformed back. The tormented hero of Victor Hugo's classic, The Hunchback of Notre Dame, *was a victim of kyphosis, a disorder which made him look "like a giant who had been broken in pieces and ill soldered together." Kyphosis accentuates the normal convex curvature of the spine's thoracic vertebrae.*

The spine of the infant, like that of the fetus, is softly **C**-shaped, its vertebrae still largely undifferentiated. The first acquired spinal curve starts to form when the infant attempts to hold up his head. At about three or four months of age, when he has mastered this, the resulting curve in the cervical vertebrae has formed. It is the least marked of the spinal curves. The most pronounced curve, one unique to humans, is the lumbar curve, a concave bend in the lower back that begins to develop when the child tries to walk. His first unbalanced steps gradually yield to a confident stride as the lumbar curve deepens and balance improves, roughly by the age of two.

The spinal column is an expressive instrument. Like a cat, which arches its back in fear, man's posture reflects his feelings. Happiness, depression, authority, submission — all can be easily read in the slope and bearing of the back. The back is no less eloquent in matters of health and sickness. Its bony axis casts an influence over the entire body. When the spine shifts, the body shifts accordingly.

In Shakespeare's portrayal of King Richard III, the angry king mocks his body and life, which has been ruled by his misshapen back. A malevolent presence in two plays, Richard's character is more twisted than his body. Shakespeare made Richard's deformities the source of his discontent and the symbol of his evil. "Why, love forswore me in my mother's womb," Richard mourns:

> She did corrupt frail nature with some bribe,
> To shrink mine arm up like a wither'd shrub;
> To make an envious mountain on my back,
> Where sits deformity to mock my body;
> To shape my legs of an unequal size;
> To disproportion me in every part,
> Like to a chaos . . .

Spinal curves, essential for the health of the back and the body, are crippling when extreme. Normally, the spine has a slight lateral curvature. But sometimes disease coerces bone or muscle out of place, accentuating the curve. The result, scoliosis, warps posture and upsets the body's balance. Lordosis is an abnormal concave curvature of the spine, usually in the lumbar region. Similarly, kyphosis is an accentuated convex curvature of the thoracic vertebrae. Commonly known as hunchback, kyphosis affects the victim's general well-being. It reduces height by forcing the spinal column into a horizontal bulge, as though it were pulled out of place by a mighty hand. It also diminishes chest capacity, restricting the work of heart and lungs.

A French sixteenth-century stained glass window depicts the biblical story of the creation of woman. Adam sleeps, unaware that he will awaken to a companion — Eve — created from one of his ribs.

It is no wonder that "backbone" has come to describe anything that provides primary stability and support. The spine is, in every sense, the backbone of the body and it directly or indirectly serves as anchor for all other bones. The ribs connect directly to the thoracic vertebrae. Twelve pairs of resilient ribbons of bone spring from the sides of these vertebrae, their heads nestled in shallow facets. The upper seven pairs of ribs, the "true" ribs, arch around the body and attach to the sternum — the breastbone — via shafts of cartilage called costal cartilage. The remaining five pairs of ribs are "false," because they link with the sternum indirectly. Costal cartilage links the upper three false ribs, which are connected, in turn, to the lowest pair of true ribs. The lower two pairs of false ribs "float," barely hovering around the sides of the body. They do not connect with the sternum at all, but terminate in cartilage connected to muscle of the abdominal wall.

"Bone of My Bones"
God, the Bible tells us, created the first woman from a rib taken from the first man as he slept. It was not until the Renaissance that the literal truth of the story was shaken. From his careful dissections, Vesalius countered the popular supposition that man had one less rib than woman. This and other notions, he found, simply proved untrue. The publication of his anatomical discoveries provoked a torrent of religious reproof. Feeling unjustly criticized, Vesalius later burned his notes in a fit of rage.

The ribs demonstrate the essence of the skeleton's protective yet flexible framework. The bony cage they form, the thorax, hugs the organs it houses. Yet the lungs must expand to do their work. Accordingly, the ribs swing outward and upward to accommodate the taking in of air.

The sternum unites the ribs in the front of the body and also carries out part of the thorax's action of give-and-take in breathing by maintaining a flexible space between two of its three main portions. Although shaped like the blade of a sword, this six-inch plate of bone acts as a shield, sheltering the heart and lungs. Like the sternum, many bones in the human body contribute to the protection of vital organs.

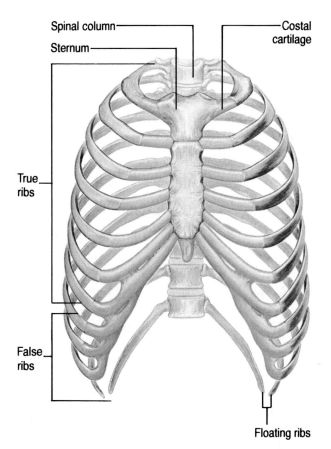

Adam's missing rib — and the twenty-three others — unite at the sternum, the breastbone, and circle the body to meet the spine's thoracic vertebrae. Their oval contour forms the thorax, a cage of bone, which protects its occupants — the heart and lungs — from injury. Ribbons of cartilage knit ribs to sternum, twisting as we inhale so that the ribs move with the expanding lungs.

Soft membrane forms a rift in the skull of this fetus, seven months old. The skull retains this soft space, one of several fontanelles, until the baby reaches eighteen months of age, when it finally hardens to bone.

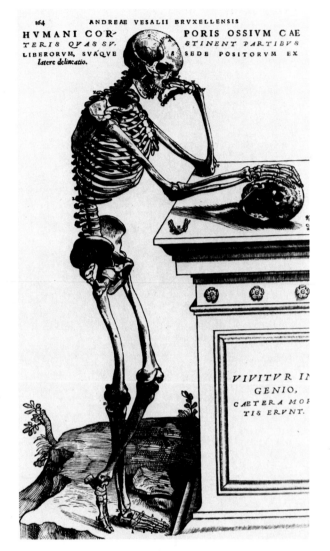

A skeleton contemplates a skull in this woodcut from Vesalius's mid-sixteenth-century studies. Resting near the skull on the corner of the tomb are bones which were rarely included in drawings of the full skeleton. The hyoid, which anchors muscles of the tongue beneath the jaw, lies below the pensive thinker's elbow. At the far right are the malleus and incus — two small bones in the middle ear — often mistaken for the artist's signature.

In the skull, protection is an overriding purpose. An enduring and unsurpassed symbol of death, the naked skull retains a chilling look of humanity, its spare countenance forever fixed with the foundations of a human face. A woodcut in Vesalius's masterwork, *De Humani Corporis Fabrica*, The Structure of the Human Body, shows a skeleton thoughtfully meditating over a skull. In a twist on a classic image, death ponders death. Some scholars suspect that this picture inspired Shakespeare's graveyard scene in *Hamlet*. If so, it makes the inscription that appeared on an early version of the print all the more relevant. "Genius lives on," it reads, "all else is mortal."

The young adventurer in Byron's epic poem of the early nineteenth century, "Childe Harold's Pilgrimage," appears in a similar stance. While wandering through Greece, Childe Harold discovers a skull lying in the dirt. He pauses to

> Look on its broken arch, its ruin'd wall,
> Its chambers desolate, and portals foul:
> Yes, this was once Ambition's airy hall,
> The dome of Thought, the palace of the Soul.
> Behold through each lack-lustre, eyeless hole,
> The gay recess of Wisdom and of Wit
> And Passion's host, that never brook'd
>   control . . .

The dome of thought is made of twenty-two bones united in the protection of the brain and major sense organs. Large at birth, compared to the rest of the body, the infant's skull is compressible. It contains soft spots between large slabs of bone that have not yet grown together. Passage through the birth canal usually squeezes the skull and elongates it slightly, but the head reassumes its natural shape within a few days.

The skull's soft spots, fontanelles, are fibrous membranes that will eventually harden and grow together until the bones meet and mesh much like the teeth of a zipper. There are six main fontanelles. The largest is a diamond-shaped region near the center of the top of the skull — an opening about the size of a quarter, but elongated. It is the last one to close, a process completed when the child is about eighteen months old.

The large skull of the infant reflects the rapid development of the brain, which matures far faster than the rest of the body. Two-year-old

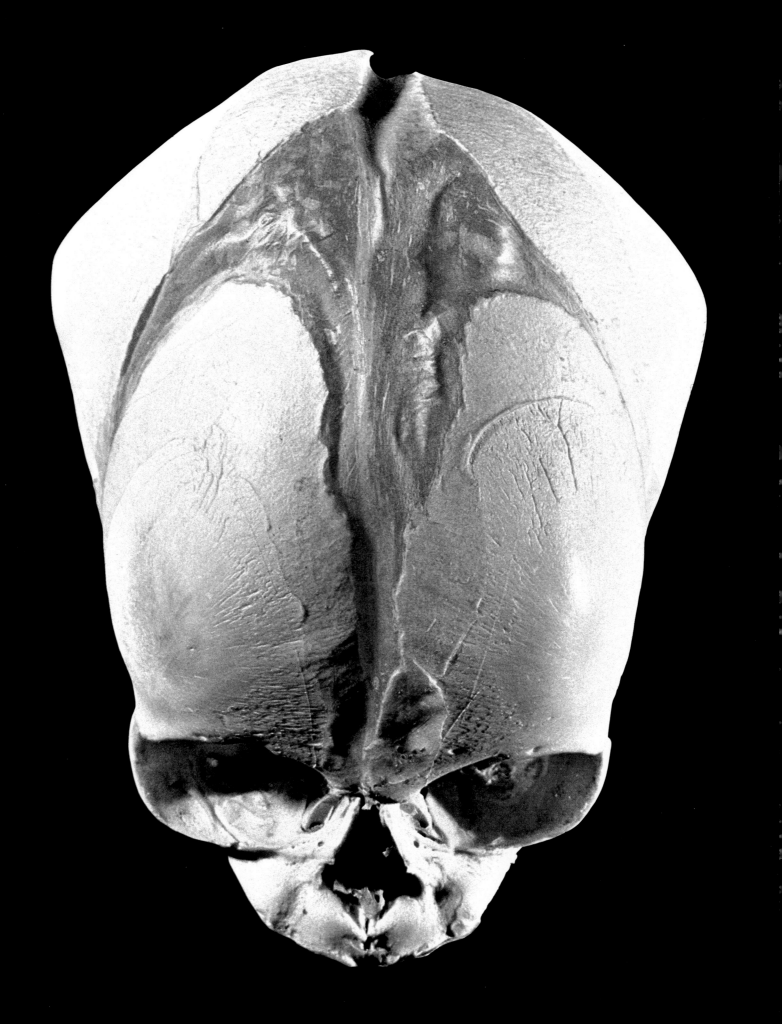

Nineteenth-century French anatomy
students studied the skull with the
aid of this lithograph, which showed
the bones separated at normally
immovable joints called sutures.
There are twenty-two bones in all.

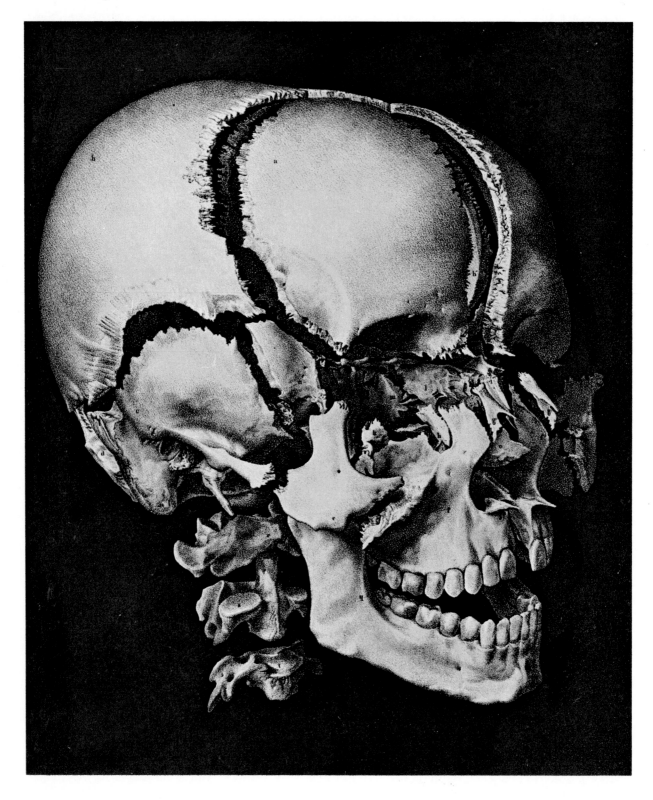

children have nearly adult-sized brains. The baby's skull, about one-fourth the size of his skeleton, echoes this advanced development. The adult skull, by comparison, is roughly one-eighth the skeleton's size. But not all of the infant's head keeps pace with its brain. The face is only about one-eighth the size of the cranium, the vault of bone resting over the brain. By adulthood, the proportions have evened out somewhat, with the face about half the cranium's size.

Facial bones undergo two pronounced bursts of development. The first major change begins with the arrival of the child's first set of teeth between the ages of six and twenty-four months, causing the face and jaws to enlarge. When the child is about six years old, his permanent teeth begin to emerge, bringing in the second major change in the facial bones. At about this time, the sinuses begin to develop. Air-filled pockets in the bones, they further enlarge the face. Puberty brings another spurt of growth in the skull overall, particularly in the face, as the sinuses enlarge.

There are four groups of sinuses, each named for the skull bone in which it lies. The frontal sinuses are located above the eyebrows. The largest sinuses, the maxillary, are in the cheekbones. Others, called ethmoid, appear in small pockets just behind the bridge of the nose. The sphenoid sinuses lie in a bone that cradles the brain. We are usually aware of our sinuses only when they become irritated. The sinuses contain an opening linked to the nasal cavity. When a cold or allergy causes their mucous membrane linings to swell, the tiny passageways are sealed off. Without free movement of air into the sinuses, pressure mounts and causes a sinus headache.

The sinuses are an important part of the skull's architecture. Their hollow chambers help lighten its weight and add to the voice's resonance. The skulls of animals, it seems, have been used to enhance man's voice. In the late 1800s, workmen who were demolishing a church in Edinburgh, Scotland, found the skulls of horses hidden behind the building's sounding board, a structure designed to project the minister's voice. The hollow skulls increased the board's sonorous effect. Other churches later used a more accessible commodity, earthenware jars, for the same purpose.

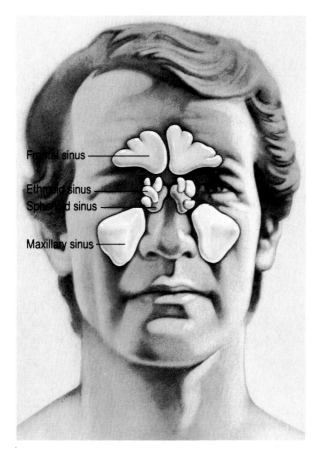

Frontal sinus
Ethmoid sinus
Sphenoid sinus
Maxillary sinus

More than a hollow vessel, the skull is an ivory fortress designed to ward off assault. Its spherical shape helps deflect blows and its bones are elastic enough to withstand all but severe blows. The fortress has carefully guarded entrances — openings behind the eyes, ears, mouth and nose, and another at its base. The large round portal at the base is the foramen magnum, the gateway from which the brainstem descends and passes into the vertebral canal as the spinal cord.

The smooth vault of skull, the cranium, is built of eight bones that interlock at immovable joints called sutures. The frontal bone curves around the skull to create the forehead and roof of the orbits, bony sockets holding the eyes. Early in this century, it was commonly believed that a large forehead indicated a large brain and, therefore, a large intellect. The assumption, however logical, was false. Man's brain size has no bearing on the degree of his intelligence.

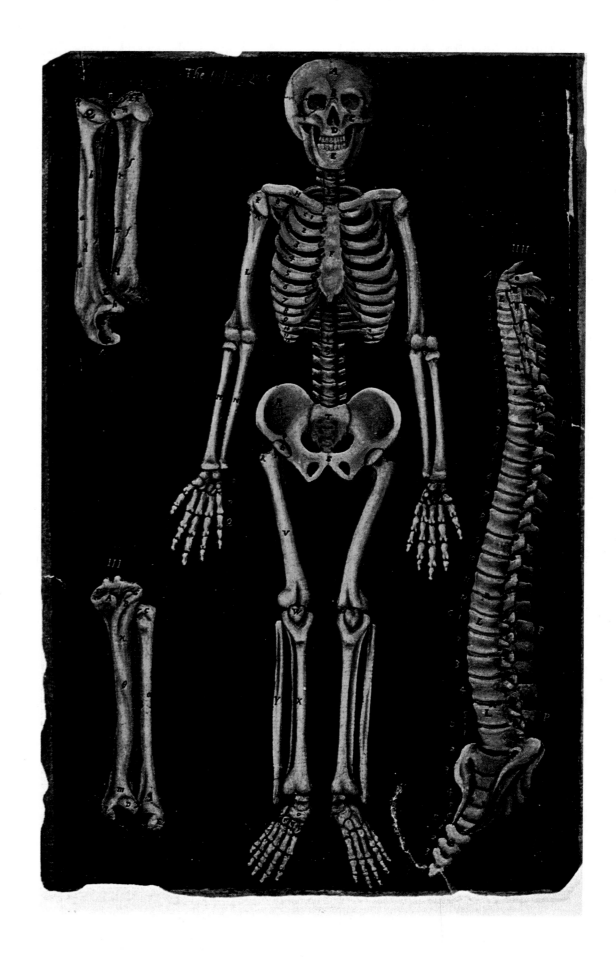

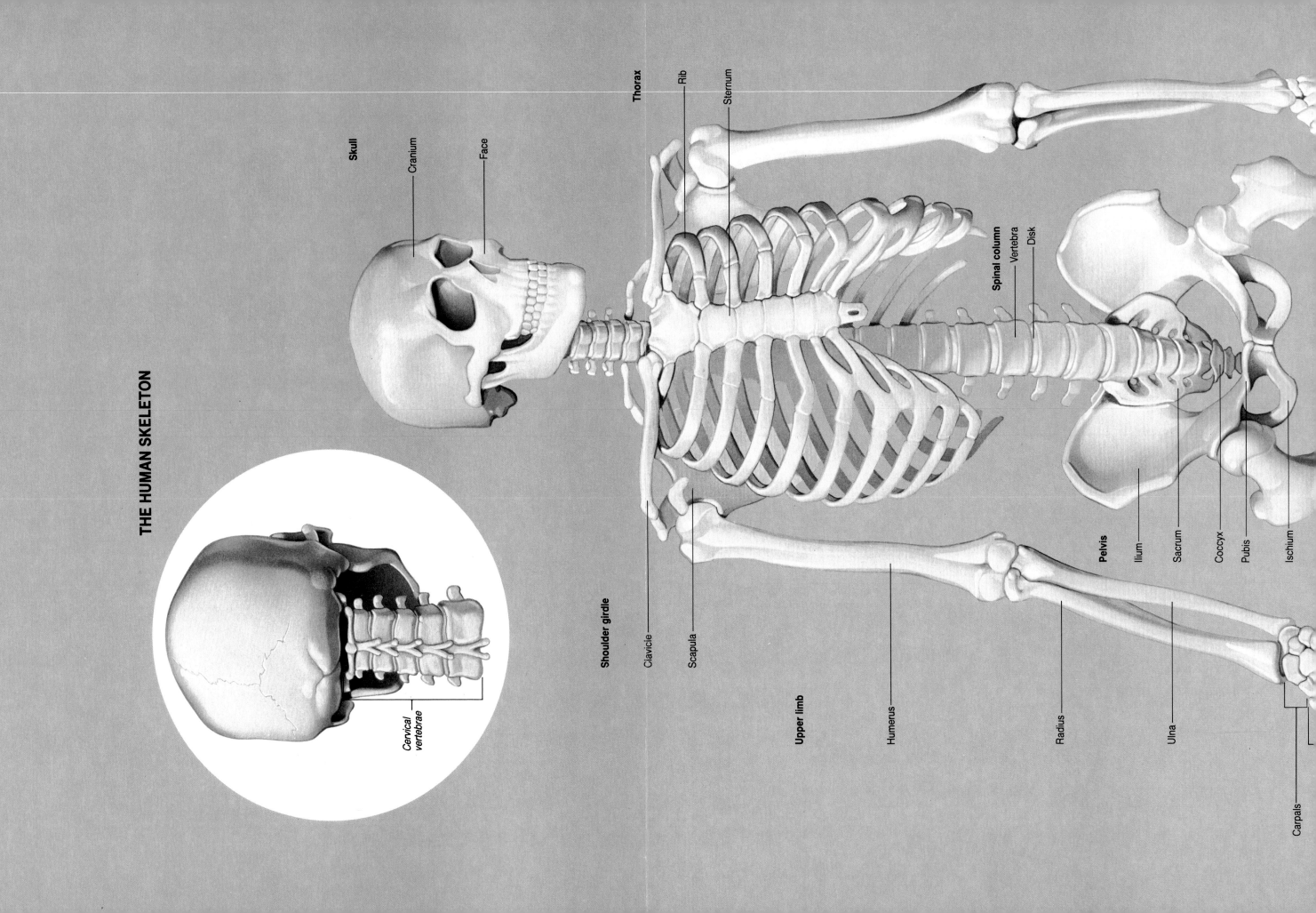

THE HUMAN SKELETON

Skull

Cranium

Face

Thorax

Rib

Sternum

Spinal column

Vertebra

Disk

Pelvis

Ilium

Sacrum

Coccyx

Pubis

Ischium

Shoulder girdle

Clavicle

Scapula

Upper limb

Humerus

Radius

Ulna

Carpals

Cervical vertebrae

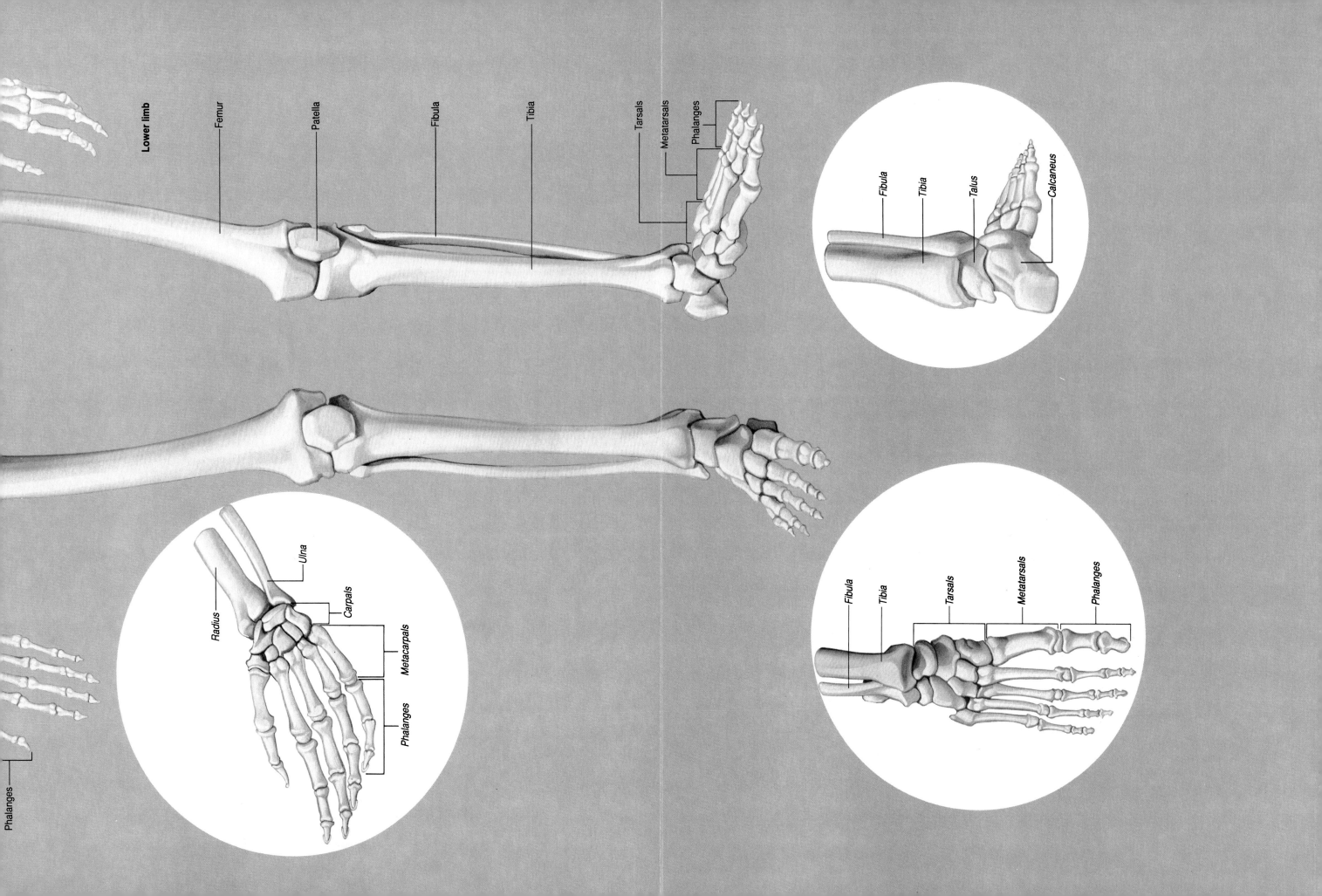

Lower limb

Femur

Patella

Fibula

Tibia

Tarsals

Metatarsals

Phalanges

Fibula

Tibia

Talus

Calcaneus

Phalanges

Radius

Ulna

Carpals

Metacarpals

Phalanges

Fibula

Tibia

Tarsals

Metatarsals

Phalanges

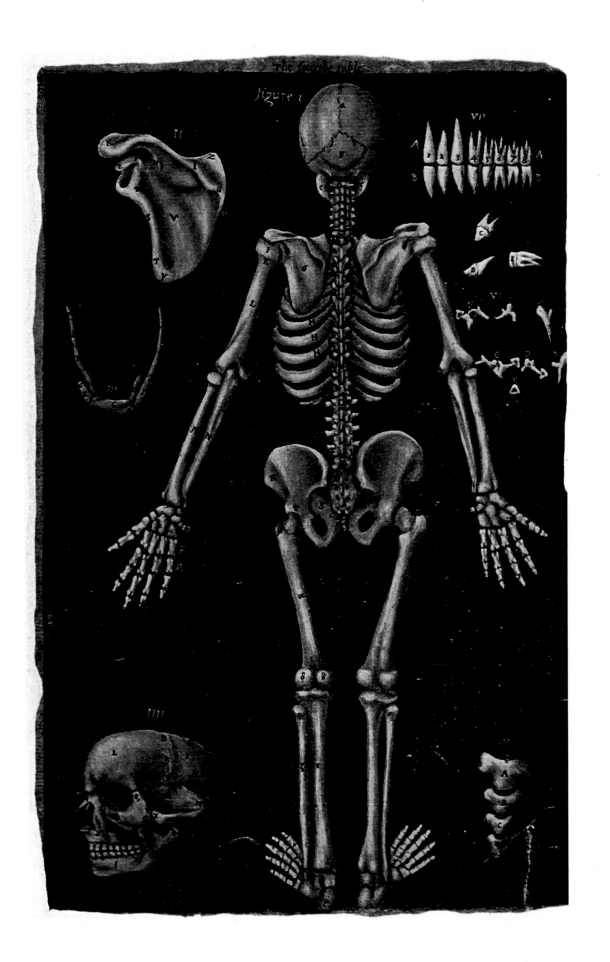
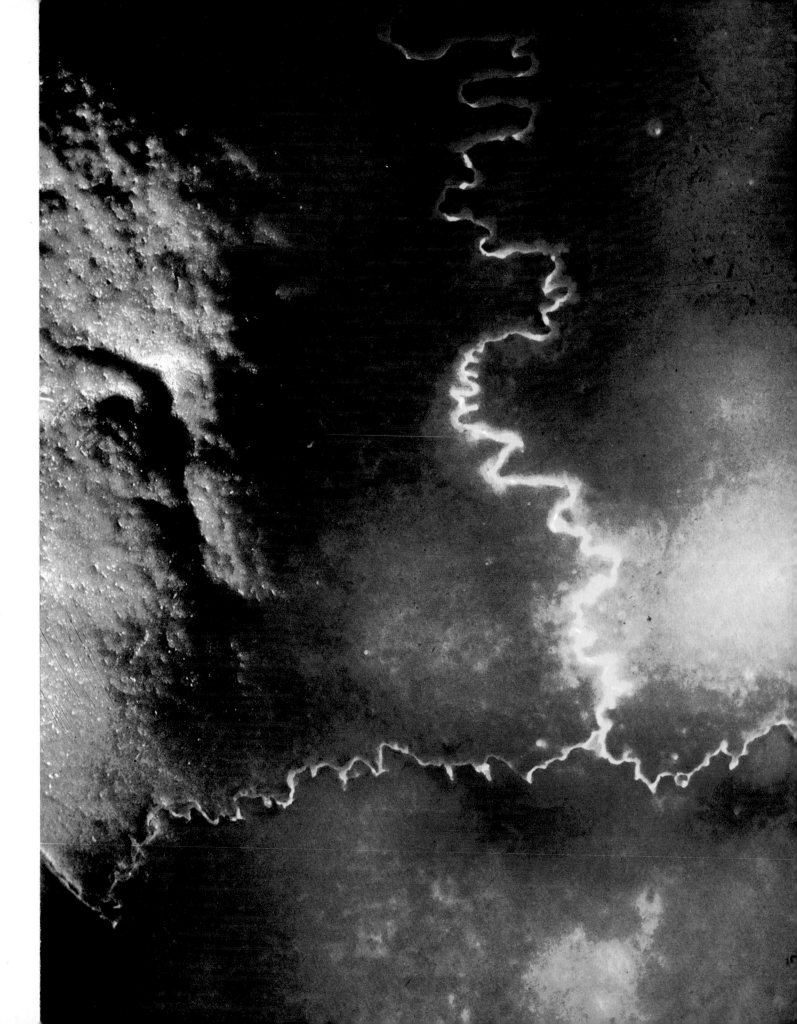

Two temporal bones, forming the sides and a portion of the skull's floor, contain canals that lead to the middle and inner ear. In the middle ear, just behind the eardrum, lie the smallest bones in the body. Each ear has a set of three bones: the malleus, incus and stapes (more commonly called the hammer, the anvil and the stirrup, from their individual appearances). The malleus is attached to the eardrum. It picks up sound vibrations from the eardrum and passes them along to the incus and stapes. This relay method intensifies the vibrations; by the time they reach the stapes, their force is about twenty times greater than in the eardrum.

Paleontologists believe the role of these tiny bones in hearing is a comparatively recent adaptation. Fish sporting early skeletons some half a billion years ago had hinged arches of small bones that supported gills. They had no jaw, no limbs and only a small opening that served for a mouth. The mouth gradually widened, crowding the gill bones. Bony arches near the mouth enlarged to form the lower jaw and other portions of the arches fused with the skull to hold the new jaw. This seemingly simple evolutionary step took millions and millions of years to complete. The newly fused bones held a tiny membrane linked to a breathing orifice — a primitive conductor of sound waves. Warm-blooded animals, man included, have six ear bones, also derived from early jaw joints.

The centerpiece of the cranial bones, the sphenoid, is bat-shaped with two large wings and two smaller "feet" extending from its central body. This "noteworthy bone," as Vesalius called it, forms much of the cranium's base and helps anchor most other cranial bones.

The Badge of Identity

Although irregularly contoured, most cranial bones convey a sense of purpose in their structure. Their exterior surfaces are smooth and curved. But their inner walls are indented and convoluted, conforming with the contours of the brain — as though the bones had been used as a mold for the brain.

Fourteen bones combine to make the human badge of identity, the face. The main bones, the

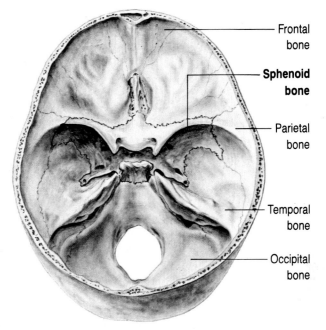

Frontal bone

Sphenoid bone

Parietal bone

Temporal bone

Occipital bone

two maxillae, act together as a keystone for other facial bones. The maxillae articulate, or link, with all but the lower jaw. They form the entire upper jaw, part of the orbits, the nasal cavity and the section of the roof of the mouth that contains the deep sockets for the upper teeth. At either side of the face, the ends of the maxillae swell forward. Here, they link with the small zygomatic bones, just below each eye. Fashion models owe their enviable facial structure to prominent curves in these, the cheekbones.

Two bones, standing side by side, shape the upper part of the nose. Most of the nose is sculpted of cartilage attached to the skull around a pear-shaped nasal opening. This opening is divided in half by the nasal septum, a partition created by bone and cartilage.

The largest facial bone and the only one in the skull with a movable joint is the mandible — the lower jaw. It is a strong curved bar of bone that holds the lower teeth. Arms of bone reach up its sides to articulate with the skull's temporal bones, creating the jaw's hingelike joint. Tucked beneath the mandible is the **U**-shaped hyoid, the only bone in the body that does not articulate with another bone. It anchors muscles, particularly those of the tongue.

Most of the bones in the body lie not in the body's central axis but in its extremities, where a wider range of movement is necessary. Arm and leg, hand and foot, are superficially similar. The number and arrangement of bones in each are strikingly alike. Yet they have functional differences that date, anthropologists believe, from the time man's ancestors first stood upright. This event ushered in structural changes.

A Constant Compromise

The pelvic girdle, the bony construction that connects the legs to the rest of the skeleton, reflects the weightier duties it inherited. Charged with bearing the weight of much of the body, helping to hold it upright and move it forward, the pelvic girdle consists of two heavy, substantial hip bones. Each hip bone is built of three once-separate bones. The broad, flaring wings at either side are the ilia. The bones of the pubis curve forward and those of the ischia create the lowermost portion of the pelvic girdle on which, according to nineteenth-century physician Oliver Wendell Holmes, "man was designed to sit and survey the works of Creation." The basinlike pelvis is formed by the hip bones and the spine's sacrum, which anchors both of them in place.

The shoulder girdle is as flexible as the pelvic girdle is stable. Its comparatively delicate construction reflects the freedom of movement granted to arms spared the burden of carrying the body's weight. The components of the shoulder girdle — two clavicles, or collarbones, and two scapulae, or shoulder blades — are loosely linked together by ligaments rather than firm joints. The shoulder girdle's attachment to the thorax is similarly fluid. It links directly with the skeleton only at the sternum. The two triangular scapulae are bound to the back by muscles, thereby freeing them to follow the movements of the arm.

Where the body needs stability, it sacrifices some freedom of movement. Where movement is granted, some stability is forsaken. The bones of the limbs reflect this constant compromise. The leg bones, in supporting the body, are much stronger than the bones of the arm, but are capable of much less movement. The largest bone in the body is the femur, or thigh bone. Long and

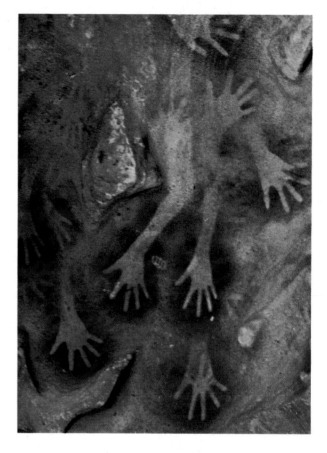

elegantly proportioned, it is also the strongest and heaviest of the body's bones. At the top of the shaft, the bone's head juts diagonally upward. Its smooth rounded surface snugly articulates with a socket in the outside of the hip bone.

Two rounded protrusions at the femur's lower end meet two dishlike surfaces on the top of the lower leg's main bone, the tibia, forming the knee joint. This is the largest joint in the body and one of the most stable. It is reinforced by the patella, or kneecap — a roughly triangular bone — which articulates with the femur. The kneecap, lodged in the tendon of the muscle which straightens the leg at the knee, increases the muscle's leverage while protecting the knee joint.

The tibia's shaft is less smoothly rounded than those of other long bones. Its surface is prominent at the front of the leg, where the strongly ridged shinbone protrudes under the skin. The other bone of the lower leg, the fibula, resembles

X-rays reveal the inner structure of the hand, mankind's most useful tool. The apparent delicacy of the twenty-seven bones in the hand and wrist belie their strength. The hand can exert great force.

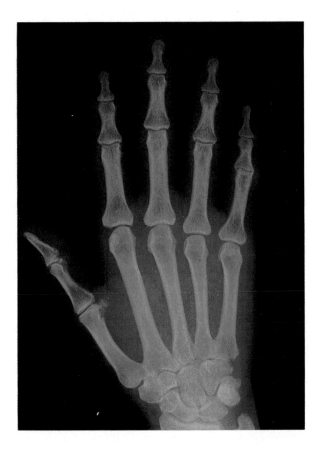

the body but gives it little stability at the joint. This is the joint in the body most in danger of dislocation. The opposite end of the humerus meets two bones of the forearm, the ulna and the radius, to create the elbow joint.

The predominant bone in the elbow is the topmost portion of the ulna, a sturdy block of bone. The head of the radius is buttonlike, rounded at the edges and flat on top. The flat top glides against the surface of the humerus and the rounded edges swivel against the ulna. This design greatly increases the movement of the arms. When the hand is held at the side of the body, palm outward, the radius and ulna run essentially parallel to each other. But when the palm is turned inward, the bones in the lower arm rotate, with the radius crossing in front of the ulna. The normal resting position of the arms and hands is somewhere in between.

Of Elegance and Might

Anthropologists believe man's earliest predecessors were quadrupeds, animals that walked on four feet. Later ancestors were tree-living animals whose four "feet" were used as hands. Only in man are the roles of hand and foot entirely separate. Sir Charles Bell, a Scottish anatomist and author of an 1834 treatise, *The Hand,* found himself awed by its design. "We must confess," he wrote, "it is in the human hand that we have the consummation of all perfection." One has only to watch a potter, a surgeon or a seamstress at work to appreciate the precision of the hand as a tool. But the hand is also eloquent. The gestures of a mime create places and objects where there are neither. Umpires and referees, whose voices become almost useless during a game, use their arms and hands to deliver judgment on a play. Angry players and coaches may in turn use a different set of signals — the same ones common to traffic jams and schoolyard fights — to comment on the official's decision. For deaf-mutes, the hands must substitute for language entirely.

The hand's great flexibility begins in the wrist, where eight carpal bones are arranged in two rows of four each. They look somewhat like small pebbles, yet their placement and articulations are precise. Beyond them, five metacarpal

the jointed side of a safety pin. Taking its name from the Latin for "buckle" or "clip," the fibula articulates with the tibia at its top and bottom. This slender, twisted bone acts mainly to anchor leg muscles. Two bony lumps at either side of the ankle, commonly called the ankle bones, are actually the prominent ends of each lower leg bone. The tibia surfaces at the inside of the ankle, the fibula on the outside.

While the construction of the arm superficially resembles that of the leg, its ability to move is far greater due to subtle differences in its structure. The humerus is the long single bone in the upper arm. A pun on its name has given us the term "funny bone." Like the femur, the humerus has a smoothly rounded head which fits into a socket joint. The socket in the scapula, however, is very shallow and much smaller than the head of the humerus. This arrangement allows the arm to move at almost every possible angle away from

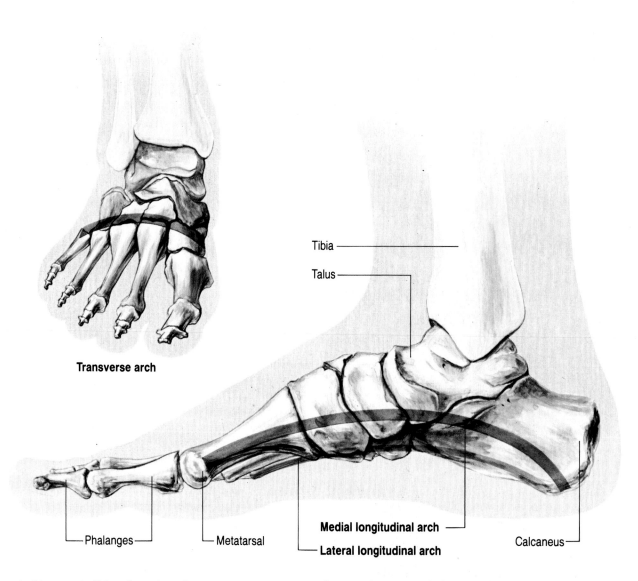

Transverse arch

Tibia

Talus

Phalanges

Metatarsal

Medial longitudinal arch

Lateral longitudinal arch

Calcaneus

Architecture is "the adaptation of form to resist force," proclaimed nineteenth-century English art critic John Ruskin. The structure of the foot does just that. An arch is the most efficient construction known for supporting weight. Arches in the foot — two running its length, another spanning its width — help provide a stable base for the body.

bones fan across the hand. Largely alike, they differ mainly in length. The special construction of the thumb allows it a greater range of motion than the other digits. With its metacarpal bone placed lower than those of the other fingers, the thumb has a sweeping range of movement across the palm. In each hand, there are fourteen bones called phalanges distributed among the fingers. The thumb has two; all other fingers have three.

It is easy to appreciate the power and poetry that are the human hand. By comparison, the foot seems a brutish thing. Yet many anatomists think the foot — not the hand — may be the most recently specialized structure. The lowly foot, amazingly resilient to the great pressures and strains it encounters, sacrificed dexterity to provide stability for the body.

The ankle has only one bone less than the wrist. But the arrangement of its seven tarsal bones — different from that of the carpal bones in the wrist — increases their weight-bearing efficiency. When the body is standing, its entire weight funnels down through one of the tarsals, the talus. There, the weight divides evenly. Half travels to the calcaneus, the heel bone, while the other half channels to the five remaining tarsal bones and into the bones that make up the arch of the foot. High-heeled shoes upset this balanced distribution by throwing most of the body's weight to the bones in the ball of the foot.

Slender metatarsal bones join in the central portion of the foot itself. These five bones articulate with the ankle's tarsals above and the bones of the toes below. They form the arch, the most important structural part of the foot. Like cathedral ceilings that soar by virtue of barrel vaulting and bridges that leap over water via great curving spans, the foot disperses body weight through three arches — two over its length, one across its width. The arches also provide the leverage necessary for locomotion.

Toes have the same number of bones as the fingers do and go by the same name, phalanges. They are also distributed similarly. The big toe has two bones and the others have three. Yet the bones of the toes have little in common with those of the fingers. Their stout shape enables them to bear weight. They are extremely sturdy. In a singular act of force, these bones help the foot push off with each stride. Firmly gripping the ground, they help us balance as we stand.

It seems a wonder that we can stand at all. The human skeleton, a tower balanced on two small feet, contradicts common-sense rules of design. No architect would dare plan such a structure except, perhaps, in fun or fantasy. The skeleton as a structure is a taunting paradox, for it succeeds beyond any architect's dreams. That this collection of bones lets us stand and move freely would lead one to suspect that the skeleton is not the exception to a rule but is the rule itself.

Chapter 3

Matters of Bone

In the seventeenth century, after conquering Okinawa, Japanese invaders confiscated all weapons and banned their manufacture. In response, the Okinawans refined a weaponless system of self-defense based on the combat techniques of Chinese monks. Known as karate, from the Japanese for "empty hand," the system evolved into a powerful fighting art.

Karate owes its success to concentration, precision and split-second timing. By comparison, Western fighting styles often seem rather crude. The karate master focuses on a small area of his opponent's body and aims to terminate his blow less than an inch beyond the target. Thus, his fist reaches top speed at the instant of contact.

Even a karate novice can quickly learn to split a block of wood. Raising his fist over such an unyielding target, he brings it down at speeds reaching forty-three feet per second, exerting a force of 675 pounds. For all the karate expert's skill, however, the feat would end in utter disaster but for the astonishing strength of human bone. Engineers measure strength in several ways. But by almost any measure, bone is among the strongest materials devised by nature. One cubic inch can withstand loads of at least 19,000 pounds — about the weight of five standard size pickup trucks. This is roughly four times the strength of concrete. Indeed, bone's resistance to loads equals that of aluminum and light steel.

Bone seems even more remarkable when one considers how light it is. The skeleton accounts for only 14 percent of total body weight, or about twenty pounds. Steel bars of comparable size would weigh four or five times as much. Bone, ounce for ounce, is actually stronger than steel and reinforced concrete. Without bone's great strength, the karate master's hand would shatter with one chop. Nevertheless, fractures can occur, because bones are often subjected to many opposing forces working in concert.

Bathed in polarized light, a slice of bone glistens with iridescent colors. Coming in two varieties, compact and cancellous, bone derives extraordinary strength by blending small crystals with threads of collagen, an elastic protein. Ounce for ounce, the fabric is stronger than steel.

57

A mere comparison to materials of brute strength, however, hardly does bone justice. As the body's internal scaffold, the skeleton must anchor flesh and tissue, what Shakespeare called "the paste and cover to our bones." Bone is the armor of man's delicate organs. Unlike a steel shaft, useless when broken, bone is living tissue that repairs itself. Bones also fit snugly at joints and transmit the power of muscle across them. Most joints are sealed for life, but must remain constantly lubricated to prevent friction from grinding them down. These combined powers transcend those of ordinary steel and concrete.

Bone derives its strength by weaving protein and mineral into a resilient fabric. Each component enhances the strengths of the other. Nearly two-thirds of bone consists of various salts, mainly complex compounds of calcium and phosphorus with traces of sodium, zinc, lead and other elements. These salts form into rod-shaped crystals, which lend bone hardness and rigidity. The remainder of bone is composed of collagen, an elastic protein which makes glue when bones are boiled. Collagen fibers are studded with the mineral crystals and wind around themselves like fibers in rope. Should the minerals in a bone be removed, the bone would become so rubbery that it could be tied in a knot, like a garden hose. Without collagen, the bone would become so brittle that it would crumble into dust. Scientists believe the mineral crystals somehow seal the bone, protecting collagen from decomposition.

Because the chemical structure of bone crystal demands calcium, a low-calcium diet weakens bone. Equally debilitating is a diet lacking phosphorus, another ingredient of bone. Describing his travels in Africa in 1785, French naturalist François Levaillant reported that cattle invaded his campsite looking for leftover bones, which they chewed. When bones were unavailable, they would gnaw on each other's horns, wood and even stones. In the 1920s, scientists learned that cattle displaying this peculiar behavior were grazing on grasses growing in low-phosphorous soil and were adjusting a deficiency in their diet by chewing phosphorus-rich bone. As soon as the scientists fed the animals dietary supplements of phosphorus, the cattle dropped the

A forgotten gladiator parries a blow with an unseen shield in this anatomical version of a Roman statue, drawn in 1812 by French doctor Jean Gilbert Salvage. Ribs armor the viscera like a bronze breastplate.

bone-chewing habit. Like cattle, people need phosphorus for bone mineral. Today, scientists have established that an average man's body contains about twenty-four ounces of phosphorus. Eighty-five percent is concentrated in bone.

Cracking Bone's Crystal

All crystals possess flaws to some extent, and bone crystals are no exception. In bone crystal, imperfections occur when magnesium replaces calcium. Calcium carbonate, the substance of limestone, can also slip into the crystal structure. The tendency of bone to pick up impurities makes radioactive substances particularly harmful to man. Like magnesium, the element radium replaces calcium in bone. Once lodged in bone tissue, radium blasts sensitive tissues with beta particles, which rip through bone like bullets, shredding everything in their path. Even minute doses of radium — less than a millionth of an ounce — can damage the genes of neighboring cells, causing them to mutate into cancer cells. Larger amounts can wreak havoc on bone marrow and disrupt blood cell manufacturing.

Eventually, radiation can cause bone tumors and leukemia. Scientists learned about radium's grim powers earlier in this century, when women factory workers were painting radium on watch dials to make the dials glow in the dark. Because the women were moistening the tips of their brushes with their lips, they were absorbing massive doses of radium. Many suffered from cancer of the jaw, arms and legs.

Like radium, the isotope strontium-90 replaces calcium in bone crystals. A deadly product of atomic explosions, strontium-90 can enter the human body through milk from cows grazed on contaminated grass. Once in the skeleton, the isotope spews high-energy beta particles, tearing apart tissue and causing tumors and leukemia.

Impurities in bone crystal also create structural problems. As a crystal's molecules pack together, slight dislocations can develop at the points of impurity. Such weak spots buckle under tension, much as cards in a deck slip over each other. These defects, called "Griffith cracks" after the British engineer who discovered them, also occur in manmade objects. In the early days of avia-

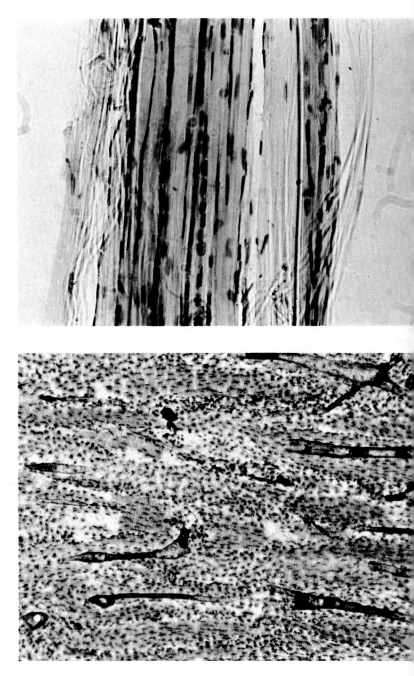

Threads of collagen studded with minute crystals make compact bone, above, one of nature's most resilient materials. Compact bone is packed in the shaft of long bones like the femur and tibia in the leg.

tion, airplanes in flight would sometimes collapse for no apparent reason. Puzzled investigators learned that Griffith cracks had spread through the airplanes' metal.

In the human body, bones are constantly stretched by tension. The action of lifting a suitcase stretches arm bones. The unique structure of bone, however, offers three powerful barriers to invading Griffith cracks. When a bone crystal splits under tension, the crack may spread throughout the crystal. The force generated by the crack triggers the start of another crack in an adjacent crystal. The second crack opens at a right angle to the original crack. When the two meet, the original crack generally stops. If the crack continues to spread, it eventually meets collagen, which bends to absorb the crack, like a pillow absorbing a sharp blow. As a final defense against Griffith cracks, bone contains millions of tiny cavities that eventually halt the spreading.

A close manmade analogue to bone is fiber glass, made of slender glass threads embedded in epoxy resin. Glass is brittle and normally covered with surface cracks. For reasons that engineers do not fully understand, however, threads of glass are usually free of surface cracks. If cracks do appear in fiber glass, they spread until meeting the epoxy resin. Like collagen, the resin is elastic and bends to accommodate the fracture.

Building on the principles that make bone and fiber glass strong, engineers have created other lightweight materials of enormous strength. The U.S. Army currently uses kevlar, a material made of tough, stretchy fibers and resin, for armor in tanks and personnel carriers. Today's military helmets are also made of kevlar, as are lightweight bulletproof vests widely used by police. Likewise, commercial aircraft manufacturers are starting to use composites of graphite fibers and resin for flaps, rudders and landing-gear doors. In building the space shuttle, NASA used a special composite of carbon fibers embedded in resin for the nose cap and wing edges. This composite holds its shape even at the searing temperatures of re-entry. Cargo-bay doors and other structures on the shuttle are made from composites of graphite fibers. NASA's designs for future spacecraft call for the increased use of composites.

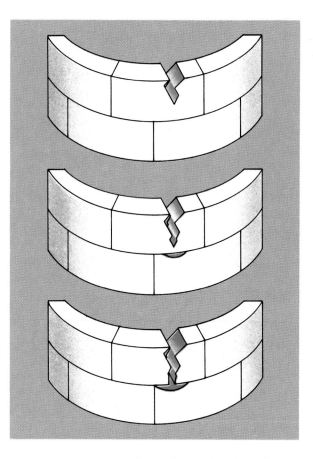

As diamond crystals split under a gem cutter's mallet, so bone crystals are cleaved by everyday pounding forces. Bone musters sophisticated stratagems to prevent cracks from spreading. One is the Cook-Gordon effect, named after the British engineers who discovered the mechanism. When a crack opens in bone crystal, the force that is generated opens a second crack in an adjacent crystal. When they meet, the cracks' march through bone is stopped.

Within bone tissue, the composite of collagen and crystal forms elaborate structures. Compact bone is concentrated in the middle of long bones like the femur, or thigh bone. A cross section of compact bone reveals an intricate pattern of concentric circles bunched together. Each circle, called a Haversian system, resembles the cross section of a tree. Corresponding to tree rings are lamellae, layers of bony tissue made of crystal-studded collagen. At the heart of large Haversian systems lie Haversian canals, pipelines containing blood vessels, lymph vessels, nerve filaments and delicate connective tissue.

Haversian Landscapes

The Haversian systems are named after Clopton Havers, the seventeenth-century Englishman who first glimpsed them through a microscope. In 1691, Havers published his findings on bone in *Osteologia Nova, or Some New Observations of Bone.* Despite his discovery, Havers was severely hampered by the inadequate powers of his microscope, and it fell upon later researchers to find more minute structures in bone.

Among these are small cavities in the lamellae called lacunae, from the Latin for "small lake." They are small indeed. A single cubic inch of compact bone contains over four million lacunae. Each harbors an osteocyte, or bone cell, which maintains mature bone tissue. It also performs the task of creating a flow of nutrients through canaliculi, minute channels in the compact bone. Radiating from each Haversian canal, the channels direct nutrients and remove waste from bone cells nestled in the lacunae. Dotting the lamellae, the lacunae are fixed in concentric rings around the Haversian canal. The first wave of channels from the Haversian canal ends at the first ring of lacunae. From lacunae in this ring, a second wave of channels radiates until it reaches the second ring of lacunae. The process is repeated until channels touch the outermost ring of lacunae, near the boundary of the Haversian system. Channels from the final ring of lacunae do not meet channels of neighboring Haversian systems. Instead, they return to the lacunae from which they originated. In this way, each Haversian system is self-contained.

Thundering booster rockets loft the space shuttle Columbia *into orbit on its maiden launch.* Columbia's *wings contain advanced composite materials resembling fiber glass that mimic the structure of bone.*

Mineral crystals — **Collagen fiber**

Haversian canal
Volkmann's canal

Periosteum

Haversian systems

Lamella

Lymph vessel **Vein** **Artery**

Lacuna

Like the landscape of some forbidden planet, layers of compact bone are organized into tree-ring structures called Haversian systems, above, named after the seventeenth-century Englishman who first discerned them. Each ring is built of fine threads of collagen embedded with mineral crystals, a sturdy blend that strengthens bone. Lying at the heart of these systems, Haversian canals are conduits for blood, lymph vessels and nerve filaments.

Rings of bone encircling a gaping portal for blood vessels encompass a Haversian system, facing page. Bunches of systems, like a fistful of wooden dowels, make compact bone. Specking the tree-ring layers, like bubbles in a lump of glass, are cavities called lacunae. These harbor osteocytes, cells that maintain bone tissue. Radiating from the central pipe are hairline channels that penetrate the surrounding tissue, bearing the traffic of vital nutrients.

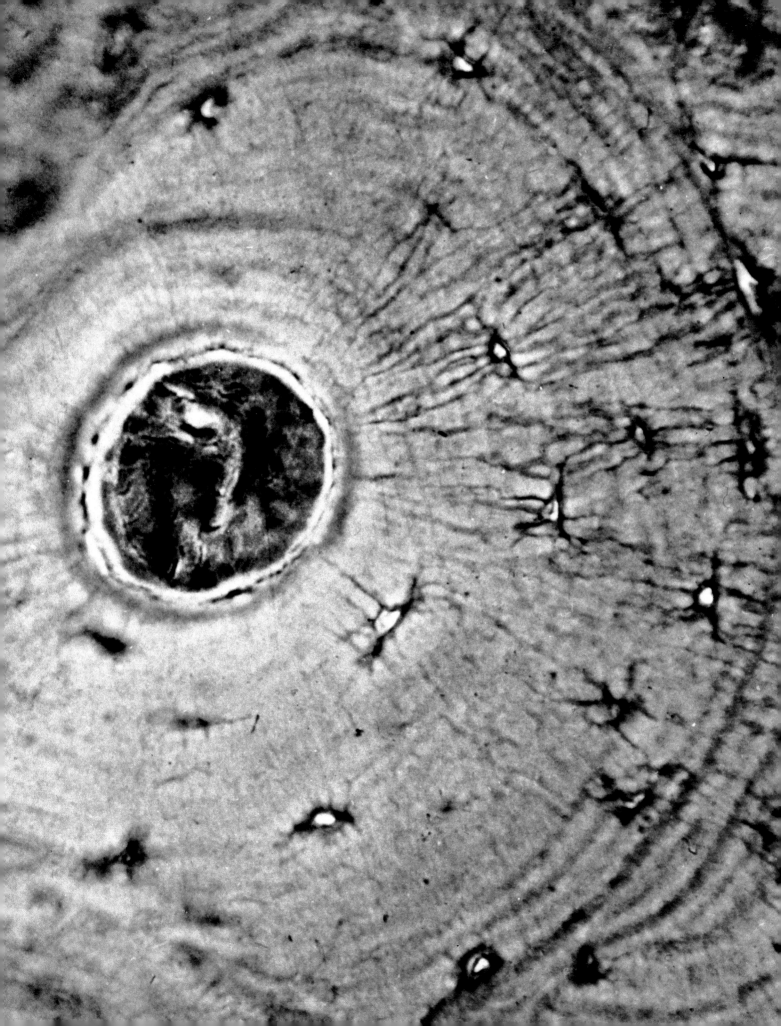

Pl. 53.

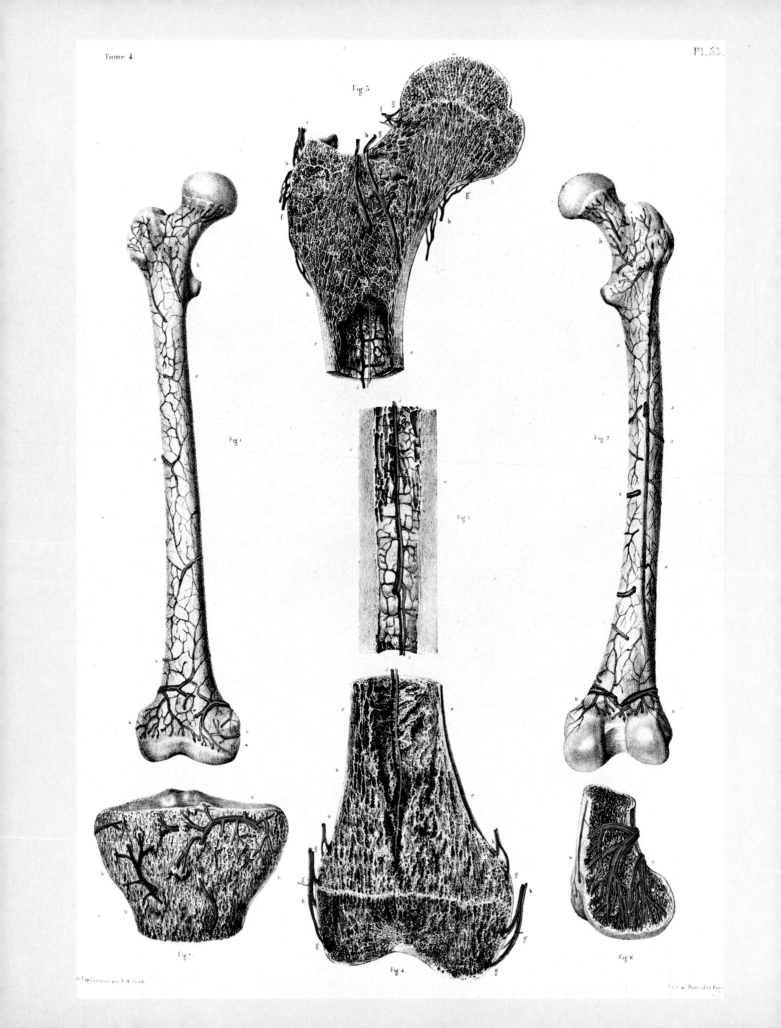

Fig. 3

Fig. 1

Fig. 2

Fig. 5

Fig. 7

Fig. 4

Fig. 6

Wrapped around the surface of compact bones,
like skin, is a thin membrane, the periosteum.
The periosteum is laced with blood vessels which
also nourish bone. When a surgeon peels the
membrane from living bone, small bleeding
points mark the spots where the blood vessels
penetrate bone tissue. When arteries threading
through Haversian canals are blocked, vessels
from the membrane can still supply local bone
cells. Thin strands from the periosteum penetrate
the bone and unite the two tissues. Other fibers
entwine with tendons, securely anchoring muscle
to bone. Because the periosteum is thick with
blood vessels and nerves, it is sensitive to injury.
Pain from fractures and bone bruises comes
mainly from periosteum damage.

Packed at the ends of long bones and lining
their hollow centers, cancellous bone, like compact bone, is built of bundles of crystal-studded
collagen. Unlike compact bone, the cancellous
variety has no Haversian systems, but instead
forms a honeycomb of struts and braces that resembles coral. These braces are the trabeculae,
joined together like the wooden beams of a scaffold. Every bone in the human body contains
both compact and cancellous bone.

Fighting Mechanical Forces

Although lighter than compact bone, cancellous
bone is designed to resist loads. The alignment of
the beams of trabeculae conform almost exactly
to lines of stress, much as the braces in a geodesic
dome are placed along stress lines. Studying this
feature of bones, engineers have fabricated artificial bones made of plastic. Laden with weights
and photographed in polarized light, the plastic's
stress lines appear as alternating dark and light
stripes. These stripes roughly duplicate the
curved patterns of the trabeculae.

Design for the accommodation of stress is vital
to the architecture of bone. In everyday use,
bones must resist different pressures. Perhaps the
most common is compression. Marble resists
compression well, making it well suited for pil-
lars and columns. Leg bones serve the same purpose. The femur, longest and strongest of human
bones, resists the compressive force of 1,200
pounds per cubic inch when we walk.

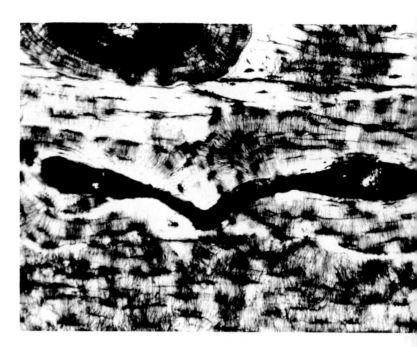

*Subterranean streams, channels
inside compact bone link Haversian
systems. Called Volkmann's canals
after the German physiologist who
discovered them, the passages allow
systems to share nutrients.*

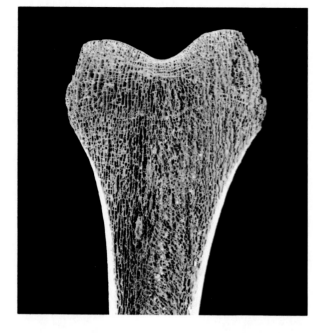

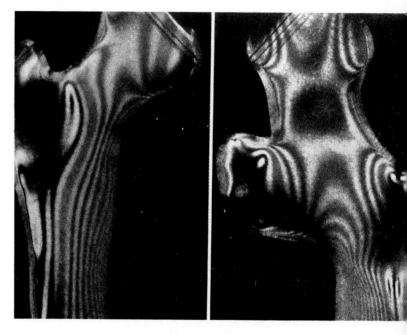

Plastic models of the femur, laden with weights and photographed in polarized light, reveal stress lines, above. The alternating stripes roughly match the arching patterns of cancellous bone, showing that the femur is designed to resist stress.

Tension, another major force, stretches bone, as arm bones are stretched when an acrobat hangs from a trapeze. Marble, by comparison, cannot withstand great tension — a fact known long ago to Callicrates and Ictinus, Greek architects who designed the Athenian Parthenon under the supervision of the sculptor Phidias in the fifth century B.C. The temple's design called for marble slabs more than twenty feet long above ceremonial entrances. Unsupported, the slabs would have buckled under their weight and the tension imposed by gravity. The architects solved the dilemma by laying iron rods under the slabs and concealing the rods with cement.

Not surprisingly, the arm bones of a gibbon, which spends much of its life swinging from trees, resist tension better than man's. But many bones in the human skeleton are constantly stressed by tension and must counter the force. The vertebrae and ribs must resist the downward drag of muscles and organs. Such a design is mimicked by sailing ships, whose masts and yardarms are stressed by tension from suspended sails and rigging.

Bones are also subject to the forces of shear. Shear stresses an object when forces are simultaneously applied from opposite directions. The act

In one hideous gulp, Greek deity
Cronus devours his child's head in
a Goya painting. However fearful,
such giants are a biomechanical
impossibility; their leg bones would
buckle under their immense weight.

of tearing a sheet of paper shears it because the hands move in opposite directions. Normally, pure shear is unusual as a force on bone. More often, shear works in concert with other forces.

Faced with these forces, bone is designed in accordance with the laws of mechanics. In every creature, including man, bone size is dictated by body weight, a principle discovered by Galileo in 1638. This phenomenon was outlined in 1926 by British geneticist J. B. S. Haldane in "On Being the Right Size," an essay which explores the relation between bone and weight. Haldane did some simple calculating, using the giants Pagan and Pope from John Bunyan's *Pilgrim's Progress*. The giants were ten times taller than the book's hero, Christian. Ten times taller would also make them ten times as wide and thick, increasing their body weight a thousandfold. Haldane concluded that, under the strain, Pope and Pagan would have broken their thigh bones on the first step.

Giants figure prominently in world mythology and folklore. The mighty Antaeus, a Libyan giant in Greek mythology, boasted that he could build a temple to Neptune with the skulls of his enemies. Antaeus drew strength from his mother Terra, the earth, and Hercules slew him only by lifting him from the ground. The Greek historian

Plutarch claimed that Antaeus's tomb was found in Africa, a huge skeleton resting within. Many legends about giants were supported by passages in the Bible. According to the Book of Genesis, giants walked the earth in Noah's time. Israelite spies sent to Canaan by Moses reported seeing giants, saying, "We were in our own sight as grasshoppers, and so we were in their sight."

Biblical evidence was cited to support the authenticity of the Cardiff giant, the most notorious hoax in American history. Carved from a huge block of gypsum into human form by a Chicago sculptor, the giant was buried on a New York farm in 1868. When the perpetrators unearthed the statue one year later, they called it a fossilized man and charged admission to view him. A minister from Syracuse wondered aloud how anyone could doubt that it was a fossil, "perhaps one of the giants mentioned in scripture." Many people agreed, betraying their ignorance of biomechanics and geometry. Without fundamental changes in bone structure, a race of giants could never exist.

Although bone size is fixed by weight, nature has found another way to economize in bone without sacrificing mechanical strength. Strain, according to the mechanical laws governing compression and tension, is concentrated at the ends of bones. Thus, a bone whose main task is to support weight has a dumbbell shape. The burgeoning end of the femur, where it meets the hip, is cancellous bone. Here the bony beams of trabeculae conform to stress lines. Midlength along the femur is compact bone. Because the middle carries less stress, however, it can be hollow with little loss of mechanical strength. The saving in weight is considerable. If the femur were a solid shaft, it would weigh 25 percent more. In birds, weight reduction is even more crucial, and some bones are almost paper thin. Such bones are often braced by internal supports, like the braces designed into airplane wings.

Abhorring wasted space as much as a vacuum, Nature has efficiently employed hollow bones by stuffing them with marrow. In the center of ribs, vertebrae, pelvic and skull bones lies red marrow, one of the most biologically active tissues in the human body. Every minute, red marrow churns

Galileo Galilei

A Renaissance Giant

While attending services at the Cathedral of Pisa in 1582, a young medical student, Galileo Galilei, watched an enormous chandelier, stirred by air currents, swaying in smooth arcs overhead. Bored by the sermon, he timed the arcs to the rhythm of his pulse. To his surprise, he noticed that no matter how wide the arc, the chandelier completed each swing in the same time. Later, he built two pendulums of equal length. He prodded one into a small arc, the other into a wider swing. Again, he found they kept time together.

The discovery was simple and revolutionary, for it enabled Dutch scientist Christiaan Huygens to invent the pendulum clock seventy-five years later. With this device, scientists could accurately measure time in experiments. It was the first of Galileo's many gifts to science. Like the breezes drifting through the cathedral, Galileo nudged the pendulum of knowledge onto a new path.

Galileo was born in 1564 into a once prosperous family fallen on hard times. His father was a professional lutanist who published a treatise on music theory. Galileo himself was an accomplished lutanist and organist. Late in life he declared that had he been free

to choose, he would have been a painter. Although he may have romantically colored the truth, he had artistic talent. The Italian painter Cigoli once said that Galileo taught him all he knew about perspective.

Pressured by his father to study medicine, Galileo lost interest quickly, especially after discovering the Greek mathematician Archimedes. "Those who read his works," Galileo declared, "realise only too clearly how inferior are all other minds." Pursuing his newfound passion, he left medical school in 1585 and soon became a mathematics professor at Padua. Original

theories on mechanics and motion enhanced his growing reputation, and his later explorations with the telescope helped advance Copernicus's vision of a heliocentric, or sun-centered, universe.

Galileo expounded his ideas on the mechanical strength of bone in *Two New Sciences*, published in 1638 after the Inquisition tried him for advocating Copernican astronomy. As a mathematician, Galileo knew that if a man doubled in height, his weight would more than double, since he would also grow in width and thickness. To support this bulk, Galileo noted that the bones would have to grow "far beyond their ordinary symmetry." He concluded that giants would defy geometric laws, unless bones were made of stronger material or were "deformed by disproportionate thickening," making these creatures "monstrously gross."

Four years after publishing *Two New Sciences*, Galileo died, and the Church refused to permit his burial in consecrated ground. Ptolemaic notions of the universe, however, gradually succumbed to Copernican theory. It is fitting that Galileo, scientific herald of a new age, also disproved the existence of giants, imaginative relics of the unenlightened past.

A hod-carrying Iraqi woman carts bricks on her head, right. The spine resists such backbreaking endeavors because disks between the vertebrae compress, below, like miniature water balloons squeezed between the hands. Powerful hydraulic forces in the spine hold the bricks aloft.

into the blood stream millions of red blood cells, white blood cells and platelets. To diagnose leukemia, doctors puncture bones to draw off red marrow for analysis. The late journalist Stewart Alsop described the procedure in *Stay of Execution*, vividly recalling that it felt like the bite of "a bloody great wasp."

Long bones are filled with yellow marrow, which consists mainly of fat cells. Yellow marrow is a strategic reserve of energy against the day when fat stores become depleted. Likewise, when blood becomes anemic, yellow marrow is quickly converted to red marrow for the manufacture of red blood cells.

The skeleton demonstrates other principles of engineering in bone design. The cheekbone and lower jaw, which together support muscle and teeth, exploit the same mechanical principle as carpenters' joists. In building construction, joists are rectangular rather than square and are laid

beneath floors with the narrow side positioned vertically. Such a configuration offers the best resistance to bending. The longer the joist, however, the weaker its resistance. The jaw and other bones that use this principle tend to be short.

Even in fine details, bones are designed to withstand stress. They are generally smooth because small surface cavities and notches weaken their structure by creating local concentrations of stress. At points where bone meets tendon, however, the bone's surface bulges with small crests. This concentrates stress, produced by muscle contraction, on the surface rather than inside bone, where resistance is weaker. It is a crucial feature in bone design, because muscles have more than enough power to snap bones in half. Such catastrophes are also prevented by the manner in which tendons are attached to bones. When a tendon, linked to a bone at a right angle, contracts, it pulls the bone and changes the angle.

The fibers on one side of the tendon are more tautly stretched than those on the other. Under these conditions, all the strain would be transmitted to the bone area stressed by the stretched fibers. However, unequal distribution of stress is prevented because the tendons weave where they meet bone and share the load.

Finally, unlike steel or concrete, bone has the power to remodel itself under stress. Scientists do not clearly understand how this mechanism works, but bone appears to have built-in sensors. The sensors constantly monitor stress and communicate their findings to cells responsible for bone growth and destruction. The cells then gradually reshape bone.

Made for Movement

The need for strength makes bones rigid. If the skeleton were cast as one solid bone, movement would be nearly impossible. In all vertebrates, including man, nature has solved this problem by dividing the skeleton into many bones and creating joints where they intersect. Joints come in an array of designs, each custom-built for the limb it serves. Lashed together by fibers of collagen called ligaments, and continuously lubricated to offset friction, joints permit movement.

Some joints allow no movement. Among these are joints binding plates of bone on the skull. Other joints allow limited movement but are less flexible than elbow and shoulder joints. The joints between the spinal vertebrae are united by disks of cartilage and other tissues, which allow some movement in several directions. Cartilage is composed mainly of collagen. Unlike collagen in bone, however, collagen in cartilage is embedded in a firm gel. This makes cartilage more flexible than bone. Cartilage also lacks blood vessels. Cartilage cells lie in lacunae, but they are nourished by fluids diffusing from nearby capillaries. Between the two pubic bones at the front of the body lies another joint united by cartilage. Late in pregnancy, the joint becomes more flexible to allow pelvic expansion.

Most joints offer far greater play. They are freewheeling connections called synovial joints, sturdy enough to hold the skeleton together while permitting a range of movements. Synovial

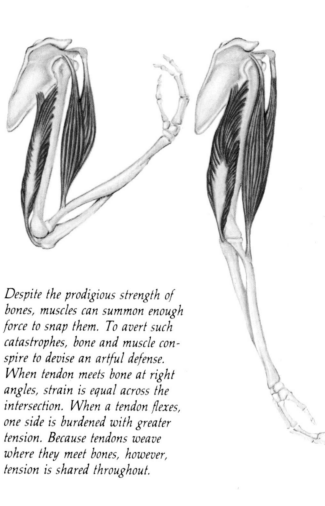

Despite the prodigious strength of bones, muscles can summon enough force to snap them. To avert such catastrophes, bone and muscle conspire to devise an artful defense. When tendon meets bone at right angles, strain is equal across the intersection. When a tendon flexes, one side is burdened with greater tension. Because tendons weave where they meet bones, however, tension is shared throughout.

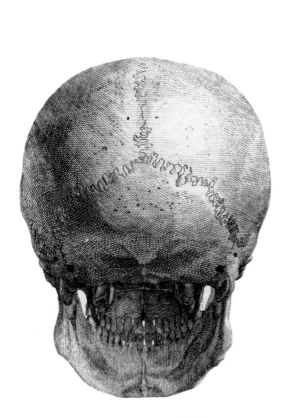

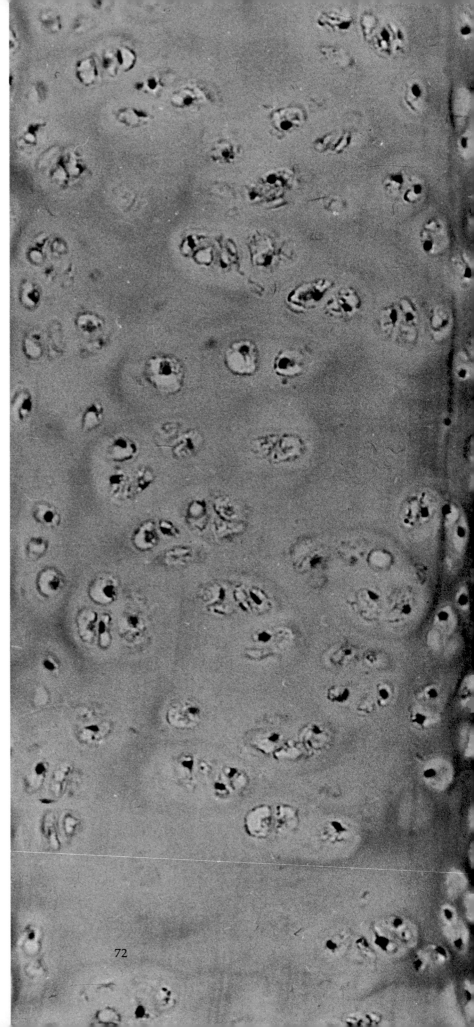

Jigsaw edges on skull plates, sutures lock together, above, holding the plates fast. At birth, the serrated joints are flexible, permitting the skull to deform for the arduous passage from the womb. At age twenty-two, the ragged sutures begin to seal — a process that continues for decades. Some joints have limited movements, including the joints of the vertebrae. These are united by pads of cartilage, right, tissue akin to bone but lacking blood vessels.

joints are fulcrums, much like the central support of a seesaw. Bones are levers, and muscles apply force. The joint between the skull and the first vertebra of the spine is the fulcrum across which muscles lift the head. Nodding one's head would be impossible without it. When a person lifts a book, the elbow joint is the fulcrum across which the biceps muscle performs the work. The mechanics of levers explains how people of identical build, size and age can have different strengths. Despite their external similarities, one person might have a biceps muscle inserted closer to the fulcrum of the elbow. This would effectively shorten the length of the lever, the bone of the forearm. In lifting a weight, this individual would be at the same disadvantage as a farmer who tries to pry a stump with a two-foot pole rather than a four-foot one.

The structure of synovial joints helps to transmit power and motion between bones. The ends of bones at synovial joints are coated with a thin layer of articular cartilage, a tough and clear variety that reduces friction and cushions the joint against jolts. If the coating is destroyed, the bones grind against each other and produce a grating noise when the joint moves. Lying between the bones, in the center of a synovial joint, is a narrow space called the joint cavity. The empty space gives the bones freedom to move. Ligaments bind bones, prevent dislocations and limit the joint's range of movement. In double-jointed persons, ligaments are unusually loose. Although generally harmless, this condition can be a sign of the hereditary disease known as Ehlers-Danlos syndrome, which causes tumors to form where the skin is injured. Because inactivity causes ligaments to tighten somewhat, runners should stretch before jogging.

Lifetime Lubrication

A sleevelike extension of the periosteum encases every synovial joint. Called the joint capsule, this structure is flexible enough to permit movement within the joint, but also has great tensile strength to prevent dislocation. Lining the joint capsule, a fine membrane secretes a lubricant called synovial fluid. A runny substance resembling egg white, synovial fluid oils the joint and contains cells that remove microorganisms and debris. Scientists have studied synovial fluid in the laboratory, where it performs poorly as a lubricant, probably because it has often been tested with metal and plastic rather than bone. Inside joints, synovial fluid is a better lubricant than almost any yet devised by engineers.

A common method of lubrication, often found in engines, is hydrodynamic lubrication. Using this system, a rotating shaft, such as a piston rod, picks up oil as it spins. This oil reduces friction by lubricating bearings where the piston rod meets the crankshaft. Scientists have shown, however, that hydrodynamic lubrication would be inadequate for human joints. Calculations show that the film of oil coating the bones would be only one-ten-millionth of a millimeter thick. The surface irregularities of articular cartilage are one thousand times greater. Under these conditions, therefore, synovial fluid could not separate the bones, and damaging friction would result.

Scientists disagree sharply on how synovial fluid lubricates joints. One theory proposes that joints are oiled by "weeping lubrication." Under an electron microscope, cartilage shows a sponge-like structure. The weeping lubrication hypothesis suggests that cartilage absorbs synovial fluid, and bones squeeze it out when they rub together. Another theory holds that "boosted lubrication" keeps the joint slippery. Under this theory, articular cartilage soaks up only small molecules in synovial fluid. Larger molecules of mucous compounds, too big to seep into cartilage, stay on the surface and oil the joint.

In addition to lubricating joints, synovial fluid eases friction between tendons and ligaments, and between tendons and bones. The fluid assigned these duties is held in small sacs called bursae. The knee alone has thirteen such sacs. Injury or excessive use of a joint can bring on bursitis, the painful inflammation of one or more bursae. Tennis elbow and housemaid's knee are both common forms of bursitis, as are bunions — bursitis of the big toe caused by ill-fitting shoes. In severe cases, doctors treat bursitis by puncturing the bursa and draining excess synovial fluid. Occasionally, in extreme cases, the bursa is removed altogether.

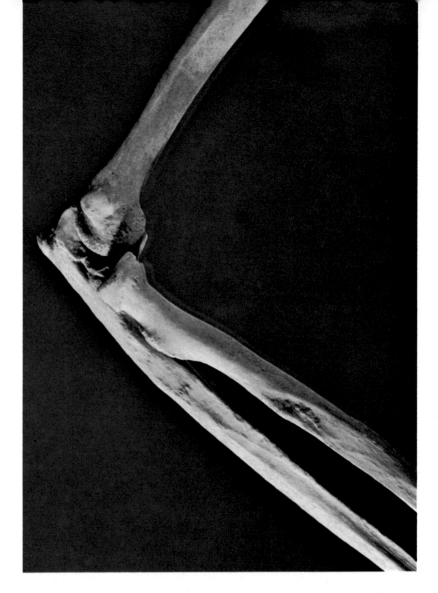

Like the hinge on a screen door, the elbow joint, right, permits movement in one direction only. Moving joints of the human machine are oiled by a runny lubricant called synovial fluid, which reduces friction.

Synovial fluid also causes a phenomenon that had long baffled scientists — knuckle cracking. Some researchers had suggested that the sound was caused by bones snapping together or by the movement of tendons. In 1971, however, three British scientists traced the popping noise to the bursting of gas bubbles in synovial fluid. Studies revealed that when a joint is stretched, pressure on the fluid is reduced, causing the bubbles to form. Eventually, the bubbles explode, releasing their energy as noise. Instead of escaping, the gas is gradually resorbed into the fluid, a process which takes fifteen minutes. An incorrigible knuckle cracker must therefore wait to entertain his audience with an encore. Generations of mothers and teachers notwithstanding, scientists do not believe that knuckle cracking seriously damages the joints.

Synovial joints come in a medley of shapes and sizes, each precisely designed for specific move-ments. The most freely moving are ball-and-socket joints. In these, the hemispheric head of one bone lodges in the hollow cavity of another. In the shoulder joint, the humerus (the bone of the upper arm) fits into the socket of the scapula (a plate of bone at the shoulder). Because the socket is shallow and the joint loose, the shoulder is the most mobile joint in the skeleton. This arrangement gives the arm a tremendous range of movements from swinging a baseball bat to reaching for a box on a shelf. This versatility also makes the joint vulnerable to dislocation, which occurs when the ball of the humerus is dislodged from the socket. If treated quickly, the condition can easily be corrected by external manipulation. Because the injury causes swelling, however, delay can make realignment difficult. If the injury remains untended, the disrupted joint capsule may heal in the wrong position. In this case, only surgery can restore the ball to the socket.

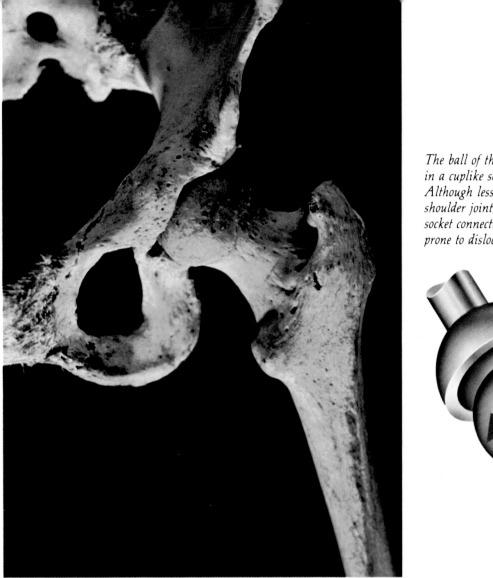

The ball of the femur lodges snugly in a cuplike socket on the hip, left. Although less mobile than the shoulder joint, another ball-and-socket connection, the hip is less prone to dislocation.

The hip joint is another ball-and-socket connection. Less mobile than the shoulder joint, the hip joint is nevertheless more stable. The ball of the femur fits tightly into a deep socket in the hip bone. A rim of cartilage lining the socket also helps to firmly grip the femur, and the ligament that unites the two bones is among the strongest in the human body.

Another basic joint design, the saddle joint connects thumb to hand. Saddle joints permit movement in two directions. Less mobile than saddle joints are hinge joints, which allow movement in only one direction, like the hinge on a screen door. Hinge joints include the elbow and finger joints. A slight variation called a condyloid joint unites the spine with the skull, permitting the head to nod. Directly below this joint on the spine is a pivot joint, which permits the head to swivel and bend. Another pivot joint, in the forearm, permits the wrist to twist.

The hinge joint of the knee, the body's largest joint, links the femur with the tibia. The femur ends in two rounded bulbs that rest on the tibia. Covering this joint is the patella, the kneecap. Unlike other hinge joints, the knee has an added dimension of complexity. The joint swivels on its axis, allowing the foot to turn from side to side. Thus, the knee is constantly rolling and gliding when we walk. To plot these shifting orientations would challenge any mathematician.

The enormous loads which the knee must bear demand a large supporting cast of ligaments and tendons for stability. Two ligaments on each side of the joint prevent it from moving too far to one side. When torn, these ligaments usually repair themselves. Strapped across the joint like bandoleers, two cruciate ligaments stop the joint from moving too far backward or forward. When these ligaments are torn, the knee is in far greater peril, for they are almost impossible to repair.

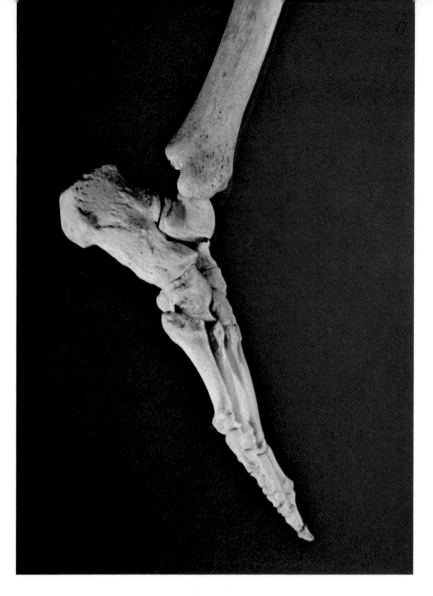

Uniting tarsal bones in the foot, between the ankle and toes, are gliding joints, right. Similar joints connect the carpal bones in the wrist, the ribs and vertebrae as well as the vertebrae themselves.

Adding further stability are two crescent-shaped wedges of cartilage called menisci, whose smooth, slippery surfaces permit the joint to glide easily. Menisci are also shock absorbers, cushioning the pounding blows the joint endures in everyday use. They are the knee's weak link; torn menisci account for 90 percent of all knee surgery. Until recently, surgeons were unaware of the vital importance of the menisci to the architecture of the knee and often removed them when they were damaged. Such operations weakened the knee's stability, making the ligaments looser and increasing friction. Although the body would grow fresh cartilage, the new tissue was thinner and less effective.

Muscles also lend support and stability to the knee. The most important is the vastus medialis, which holds the patella in place. The hamstring muscle — nemesis of many athletes — lies at the back of the thigh, enabling the knee to flex and preventing the lower leg from slipping forward. Muscles stretching along the front of the thigh control knee extension.

Knee specialists disagree on the adequacy of the joint. Some extoll its virtues, others condemn it as the body's most poorly constructed joint. According to Los Angeles orthopedist Robert Kerlan, the knee joint has remained unchanged for millions of years and is well suited for simple actions. Kerlan notes, however, that the human body "is simply not constructed for the games men play today." Beverly Hills knee surgeon Jack Kriegsman agrees: "The functions for which the knee was designed — walking, sitting and running — make the knee highly efficient. The trouble comes from sports that make unusual demands on the joint, especially when twisting and pressure are applied at the same time."

Even the most innocuous of sports often stresses the knee beyond its limits. Jogging, espe-

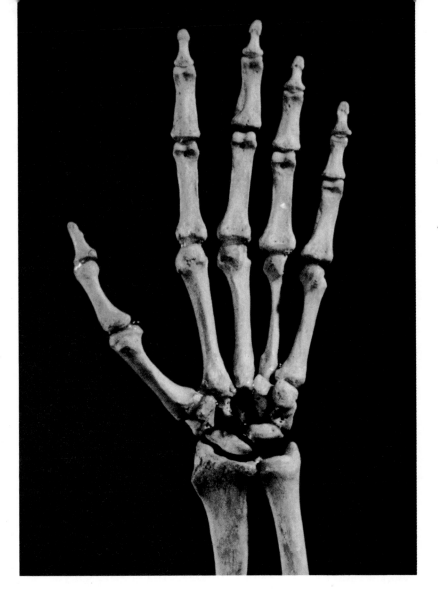

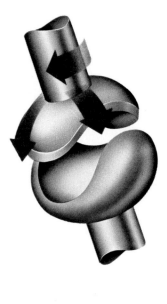

A saddle joint, permitting movement in two directions, unites thumb and hand, left. The joint, with a long, muscular thumb, permits man a grip unique among primates — crucial in his evolutionary progress.

cially on pavement, can pound cartilage into dust. Jogging injuries to the knee have become so common that orthopedists in Los Angeles have named one popular running route "The Street of the Wounded Knee."

Perhaps the most common jogging injury is runner's knee, a catchall term for knee pain that occurs when the tibia rubs against the kneecap. One cause of the ailment is flat feet. The absence of arches makes the knee turn inward so that the kneecap rubs against the femur. Until 1970, doctors treated runner's knee either with cortisone injections or by surgically shaving the back of the kneecap. In that year, George Sheehan, a physician and world-record holder in the mile for men over the age of fifty, published a paper describing how he corrected the problem in one long-distance runner by building special inserts into running shoes. Sheehan stumbled on this treatment when he noticed that his own knee

hurt when he ran with traffic, but not against traffic. Eventually, Sheehan realized the slope of the road was causing his left foot to land high. As a result, he was landing on a part of his foot that caused the kneecap to rub against the femur. Many athletes now wear Sheehan's special shoe inserts, which are called orthotics.

Knee injuries are also common in skiing accidents. Although the sport once produced twisted ankles and broken legs, later generations of high-necked ski boots have shifted twisting forces to the knee. Likewise, the twisting required to smash a tennis ball from the baseline can easily rip a meniscus. Anyone seeking refuge in bicycling as a safe sport will find that pumping up steep hills puts intense pressure on the kneecap. In extreme cases, the underside of the cap starts to fray. Even surfing takes its toll. Kneeling on a surfboard creates a hard bump under the kneecap called surfer's knob.

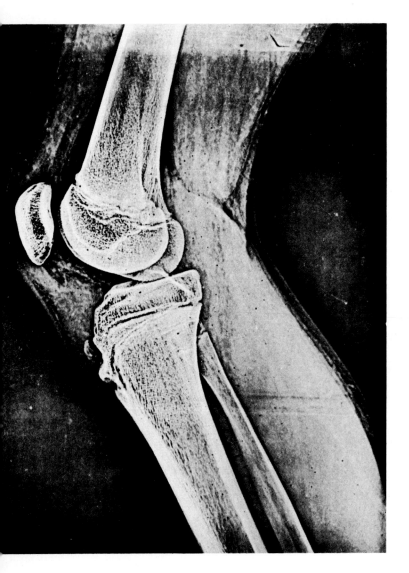

A bent knee reveals architectural intricacies to an enhanced X-ray. A complex joint, the knee glides in two planes. Coating the ends of the femur and tibia, which intersect at the joint, are thin layers of cartilage to ease the moving bones by reducing friction. Floating in the tissue beside the joint is the kneecap, which shields the joint while boosting its leverage. Tough fibers of collagen called ligaments bind the entire structure together.

Not surprisingly, aggressive contact sports place the knee in greater jeopardy. Hockey player Derek Sanderson recently explained why hockey can be so dangerous to the fragile joint: "Once you make a cut, your body is committed to where your feet go. It's impossible to get your foot out of that rut. If you come in contact with another body going twenty-five miles an hour, your knee is automatically going to go. You'll tear all the ligaments and cartilages in the knee."

That is exactly what happened to Pat Stapleton of the Chicago Black Hawks, when he crashed into a goal post at top speed. A doctor immediately put Stapleton's leg in a cast for six weeks. When the cast was removed, the leg was frozen in a bent position, and there was some doubt that Stapleton would ever walk again. Finally, another doctor instructed the Black Hawks' trainer to sit on the leg while it was propped on a chair. After several weeks, the leg straightened, and Stapleton underwent knee surgery. The trainer later said, "It's a medical miracle that Pat was able to play hockey again." Other players have been less fortunate. Bobby Orr, whose amazing feats propelled the Boston Bruins to championship a decade ago, underwent many operations before knee injuries forced an early retirement.

Probably no sport is more perilous to the knee than football. The average National Football League career is 4.6 years, largely due to knee injuries. During the 1960s, 70 percent of professional players had knee surgery before the age of twenty-six, including nearly every quarterback. In recent years, artificial turf has aggravated the problem. The main cause, however, is the sport itself, which demands that large men armored with pads and helmets charge each other, aiming for low tackles. O. J. Simpson, whose career was ended prematurely by knee injuries, admitted recently, "I used to say only bad football players get hurt. Now I say it's a violent game." Other football greats whose playing days were cut short by knee damage are Gale Sayers, Joe Namath, Bubba Smith and Dick Butkus.

Considering what might have happened, New York Jets running back Emerson Boozer was lucky. As a rookie in 1966, Boozer scored thirteen touchdowns in six games and was threatening to

demolish Gale Sayers's scoring record. In a game
against the Kansas City Chiefs, Boozer was tack-
led after making a spectacular catch. The damage
to his knee was colossal — cartilage was fractured, ligaments torn and his kneecap dislocated.
Even thigh muscles were torn. Immediately
flown back to New York, Boozer underwent surgery at midnight. His knee was completely re-
built. The operation was successful, and Boozer
played football for another ten years.

Among football players, a former linebacker
for Kansas City named E. J. Holub is legendary.
Over his career, Holub's knees were opened
twelve times by surgeons. The scars on his knees
are so dense that a friend said he looked as if he
had lost a sword fight with a midget. More operations — possibly including an artificial knee
transplant — are planned. Today, Holub runs a
37,000-acre ranch in Oklahoma, and says off-
handedly of the prospect of getting an artificial
knee, "When the time comes and they got something handy, we'll do it. It ain't no big deal."
Considering the prosthetic miracles that surgeons
now perform, Holub may be right. But whatever
type of artificial knee he eventually gets, one
thing is certain. No manmade device can replace
the extraordinary joint nature provides.

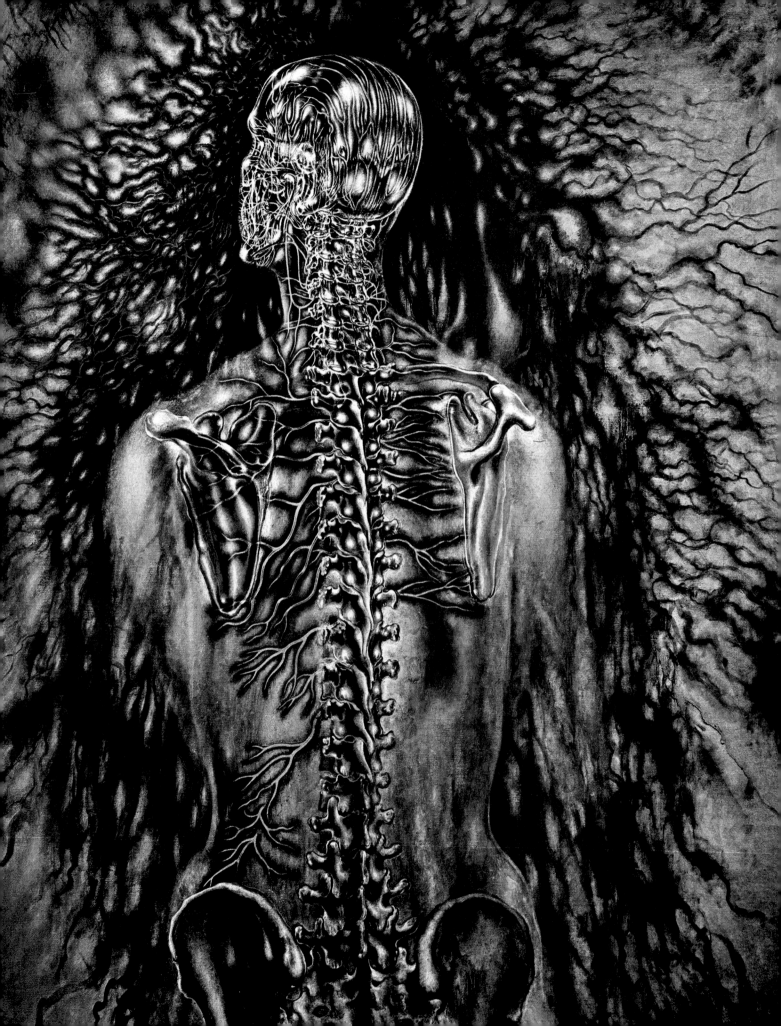

Chapter 4

Growth and Renewal

When John Belchier, a London surgeon, dined one evening in 1736 with a friend, he noticed something unusual about the pork he was served. The bones left on his plate were a deep red rather than the expected white. Questioning his host, whose business was the dyeing and printing of cloth, Belchier discovered that the dyer fed his hogs grain soaked with madder, a red dye made from a plant root used in coloring fabric. Intrigued by this accidental discovery, Belchier began to feed madder to his own chickens. When he later examined them, he found that the chicken bones had turned red, just like those of his friend's hogs.

Earlier experimenters had learned that bone grew, but the way in which it added to itself remained a mystery. When Henri-Louis Duhamel, a Frenchman who had abandoned the study of law in favor of science, heard of Belchier's observation, he determined to conduct his own experiments. Between 1739 and 1743, he dyed the food of pigs, chickens, turkeys and pigeons and found that their bones became red only on the outermost layers, the portions formed most recently. Under the layer of red, the bone was white. Next, Duhamel fed his animals madder for a while, put them back on their regular diet, then gave them madder once again. When he dissected the animals to see the effect of his experiment, he found that the bones were striped red and white. It seemed clear to Duhamel that bone grew by the addition of layers. Bones somehow had an outer wrapping that deposited the hard substance of bone, causing the skeleton to grow.

It took many years for Duhamel's appealingly simple theory to be justified. Scientists in our own century have verified that bone grows by slow accretion, gradually adding layer upon layer until adulthood. They now know the skeleton is not rigid or unalterable. It changes constantly, growing and repairing itself just like the rest of

Seeking to portray the interior landscapes of man, Russian-born painter Pavel Tchelitchew transformed the body into a vision of mystical wonder. His image of the skeleton glows with radiant energy, emitting currents that seem charged with electricity — a property of bones that scientists have confirmed.

The embryonic skeleton develops through ossification, eventually replacing soft cartilage with true bone. Cells spread from a cluster, or ossification center, in a cross section of a seven-week-old arm bone.

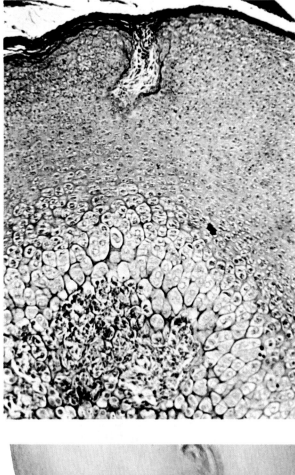

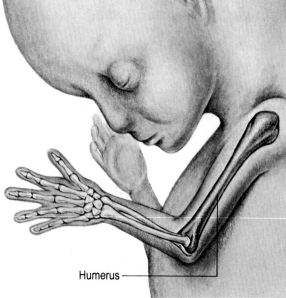

Humerus

the body. Perhaps even more than other organs of the body, it is subject to the inevitable demands of age. Furnishing the body's source of support and movement, the skeleton deteriorates in later years from the resilience of its youth.

The Making of Bone

Through the process of ossification, the bones develop and harden. Examination of embryos has shown that a skeletal outline is apparent in about the fifth week of pregnancy. For most people, ossification continues until about the age of twenty-five, by which time bone has replaced cartilage, ending further growth. Occasionally, disease or other afflictions stunt the development of the skeleton, preventing the bones from reaching complete ossification. Henri de Toulouse-Lautrec, French painter of the late nineteenth century, had a bone disorder that impaired the normal ossification of his skeleton. He was less than five feet tall and had unusually short arms and legs. His head was large, with a receding chin. The sutures of his skull never closed fully, and when he suffered broken legs as a boy, the fractures healed slowly. In recent years, French researchers have suggested Toulouse-Lautrec had pyknodysostosis, a rare bone disease that can cause incomplete bone growth and dwarfism. This disorder is most prevalent among children whose parents are related by blood. Toulouse-Lautrec's parents were first cousins.

Most of the developing bones in a human fetus are first evident in a flexible skeleton that consists not of solid bones but of cartilage, a tough gristlelike substance the color of milk glass. The cartilage is necessary because the fetus grows rapidly and is not required to support any weight. Gradually, the cartilaginous model — consisting of limbs, vertebrae and ribs — is replaced by bone through a process called endochondral ossification. Other bones, particularly the broad, flat bones of the head, develop by means of intramembranous ossification, a process not requiring a temporary model of cartilage.

The great majority of the 206 bones in the body are formed by endochondral ossification. Surrounding the cartilage model in the fetus is a thin membrane called the perichondrium. Cells lying just beneath this layer begin to form bone around the middle of a cartilage shaft. This initial segment of bone becomes the periosteum, which continues to deposit new bone material and so increases the thickness of long bones layer by layer. Inside this solid matrix, a hollow opening, the medullary cavity, remains. Yellow marrow, a fatty connective tissue, will later fill the cavity.

Between the sixth and eighth weeks of pregnancy, the fetus's bones start to grow quickly and do not stop until the teen years. By expanding at each end, bones of the young skeleton grow in length. The site of this growth is the epiphyseal plate, a layer of cartilage between the epiphysis, or end of a long bone, and the already calcified shaft, the diaphysis. Cartilage cells, created by mitosis, or division, are deposited just beyond the separation between the epiphyseal plate and the diaphysis, growing away from the freshly hardened bone. As blood vessels and bone cells approach the epiphyseal plate, the cartilage cells are destroyed and replaced by fresh bone tissue. Cannibalistic as it may seem, this is the only way our bones can increase in length. The newly deposited cartilage cells push the epiphyseal plate farther from the center of the shaft, staying one step ahead of the encroaching bone until the body has attained its full height, around sixteen years of age in women, seventeen or eighteen in men. After that time, cartilage cells at the epiphyseal plate stop dividing and are slowly taken over by bone. Ossification ends around the age of twenty-five, when the breastbone is fully developed. All that remains of the epiphyseal plates is a vestigial line.

The skeleton's second type of bone formation, intramembranous, begins by the fifth or sixth week of pregnancy. Cells from the mesenchyme, part of the middle layer of the embryo, cluster together in a fibrous membrane in what scientists call a center of ossification. These clustered cells are osteoblasts. They secrete compounds known technically as mucopolysaccharides but more commonly as cement substance. Tough collagen fibers are also secreted by osteoblasts. The fibers, becoming embedded in the cement substance, unite in a reinforced matrix. The matrix hardens by calcification when mineral salts, mostly those

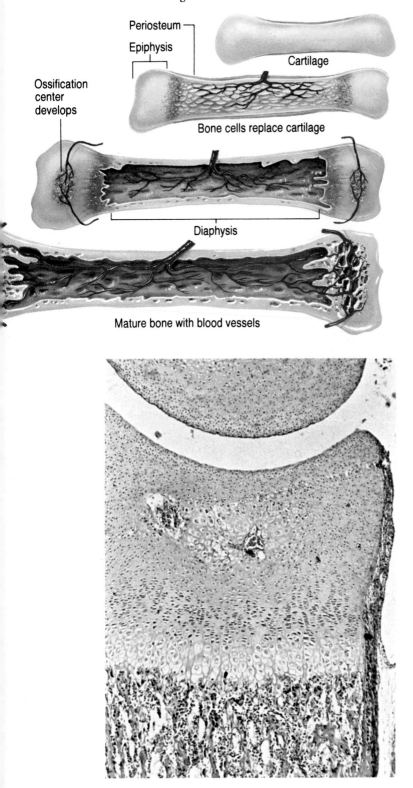

Periosteum

Epiphysis

Cartilage

Ossification center develops

Bone cells replace cartilage

Diaphysis

Mature bone with blood vessels

of calcium, are deposited. In the interior of most bones, the sticky cement substance hardens into interwoven fibers of bone called trabeculae.

As more and more trabeculae form in other embryonic ossification centers, they meet in a latticed network characteristic of the spongelike interior of cancellous bone, predominant in the broad, heavy bones of the body. The other principal kind of bone is compact bone, which makes up the shaft of long bones in the arms, legs and other parts of the body. Both cancellous and compact bone develop from the two types of ossification, intramembranous and endochondral. Every bone in the body contains both cancellous and compact tissue. Cross sections of compact bone show that it is tightly packed with concentric layers, much like the annual growth rings of trees. Bone cells lie inside these rings, amid a complicated system of canals and other openings that carry blood vessels through bone.

If the development of the skeleton is a near-perfect example of engineering, many carefully calibrated contributions make it so. Besides genes, skeletal growth is regulated by secretions from the pituitary gland at the base of the brain, the thyroid gland in the neck, the adrenal glands near the kidneys and the sex glands. Proper nutrition strengthens bone, particularly when adequate amounts of vitamins A, C and D are present. Pressure on the ends of bones, whether from cartilage, muscular development or injury, can also affect growth. Ligaments, which connect bone to bone and aid in movement, also transmit electrochemical signals to enhance bone growth.

Often overlooked is the role that emotions play in physical growth. A number of scientists have shown that emotional stress in pregnant women can have a detrimental effect on fetal development. Children raised in emotionally unsettling circumstances are, on the whole, smaller in stature than those from more stable households.

Molding itself in precise form as it grows, the skeleton adapts to its own growth and to the environment around it. The thorax, reaching from the neck to the diaphragm, loses its youthful roundness and takes on a more elliptical shape; the pelvis becomes more prominent, wider in women than in men. As children grow and be-

John Hunter

Pursuing Nature's Game

Before Charles Byrne died in 1783, he stipulated in his will that his body be placed in a lead casket and sunk in the Irish Channel. Only twenty-two years old at the time of his death, Byrne was a widely known Irish giant rumored to be eight feet tall. Eager to acquire any rare anatomical specimen, London surgeon John Hunter offered Byrne's pallbearers 500 pounds — a large sum for the time — to relinquish the body. His ploy succeeded. To this day Byrne's skeleton can be seen in the Hunterian Museum in London.

Going out of his way to find bizarre forms of life was not unusual for Hunter, a man of wide-ranging scientific interests. He populated his modest estate with birds, fish, deer, leopards and a bull presented by Queen Charlotte, wife of George III. Fascinated by nature's variety, he studied simple sea animals, the sense of hearing in fish and the development of antlers in deer. He even experimented on himself, observing the healing of his Achilles tendon after he had ruptured it while dancing. When he died in 1793 at the age of sixty-five, Hunter was England's foremost surgeon, and his teachings influenced an entire generation of British and American physicians.

Little in his early life, however, foretold his future accomplishments. An aimless young man, he was a poor student. He quit an apprenticeship to a cabinet-maker because he did not like the work. At the age of twenty-one, with no better prospects, he moved from his native Scotland to London, where he worked as an assistant in the dissecting laboratory of his brother William, a respected teacher of anatomy. He enrolled at Oxford University in 1755, but after less than two months he left. "They tried to make an old woman of me," he said later. "They wanted to stuff me with Greek and Latin at the university, but these schemes I cracked like so many vermin as they came before me." Nevertheless, he earned a reputation as an able student of anatomy. The Royal Society

of London, England's most prestigious scientific body, recognized Hunter's work by electing him a Fellow in 1767. Among the professional and honorific posts he held was that of surgeon-extraordinary to King George III.

At a time when most people — including physicians — believed bones to be no more alive than stone, Hunter taught that the skeleton was made of complex living tissue, constantly renewing itself. He was among the first to realize that movement and moderate pressure could help heal bone fractures. A patient's broken femur healed after Hunter directed him "to walk upon crutches and to press as much on the broken thigh as the state of the parts would admit." Describing the continual remodeling of bone, he wrote that new tissue "deposited in an old bone is to make up for the waste that is daily going on in it." He also understood how the skeleton developed in the womb from simple origins of membrane or cartilage. Of the growth of bones from an embryonic bud, he wrote, "The gradations of this change are beautiful." A skilled surgeon, an innovative anatomist, John Hunter never tired of the joy of discovery, of the beauty inherent in life's changing forms.

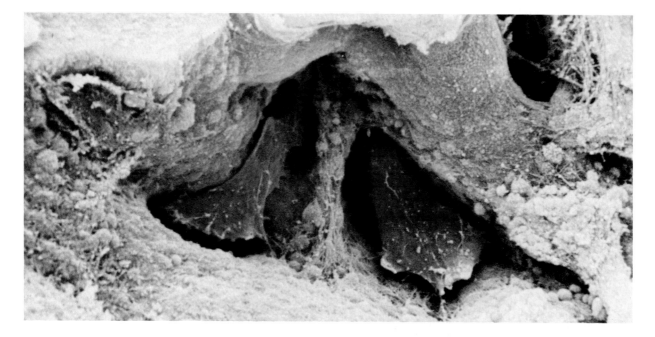

come comfortable with their upright posture, the vertebral column acquires two more curves than it had at birth, the legs become proportionately longer and the trunk shorter.

Destruction and Restoration

All parts of the body are continually in flux, and the skeleton is no exception. New cells develop, as in the skin and other organs, to take the place of older cells that have exhausted their usefulness. In the 1760s, John Hunter, a prominent British physician, was the first to discover that bone renews itself. When Hunter performed the experiments of Belchier and Duhamel by feeding madder to animals, he too found that new bone in the animals was red. But when he delayed killing an animal for several weeks after he had stopped feeding it madder, the red bone disappeared. Through natural destruction and growth, old bone cells had been eaten away and replaced. This steady process continues throughout life. Bone remodels itself to remove worn cells and to maintain the proper amount of calcium in the blood. Resorption, this breaking down of bone, occurs at different rates throughout the body. Parts of a bone's shaft, though, consisting of compact bone, may never be remodeled.

The workhorses in the metabolism of bone are two kinds of bone cells: the osteoblasts, which manufacture new bone tissue, and osteoclasts, which consume old and worn bone matter. The osteoclasts, unusually large cells having several nuclei, live only a short time as they play their destructive but necessary role in the metabolism of bone. Early in the development of the skeleton, they help bones grow to the proper length. Osteoclasts are instrumental in the shaping of bones after growth has ended and in creating the medullary cavity that runs through the middle of long bones. During resorption of mature bones, osteoclasts are active on less than 1 percent of the bone surfaces at any given time.

Developing from cells carried in the blood stream, osteoclasts are driven to action by secretions of a hormone from the parathyroid glands. Osteoclasts usually act together in clusters, burrowing minute tunnels in old bone tissue. They invade bone by sending out tiny projections called villi. The villi secrete enzymes and acids that chemically break down the bone matrix. The osteoclasts then crumble and digest the matrix, eventually depositing the matrix's minerals in the blood stream. In comparison to other body cells, which may survive for months or even several

An early explorer to eastern Burma wrote of the "giraffe women" of the Padaung tribe. Following tribal custom, a medicine man places a brass rod around the neck of a girl when she is about five years old, tightening it each year into loops. The coiled rings around this woman's neck weigh about twenty pounds. Because of pressure on the vertebral column, her neck appears to have grown longer. X-rays would reveal, however, that her chest was pushed lower, her collarbone and ribs sharply bent. With her head held rigid by the rings, she can drink only through a straw. Leg coils weighing thirty pounds compel the Padaung women to waddle and to sit with their legs extended.

years, osteoclasts are short-lived, existing for no more than three weeks. Scientists still know very little about the life cycle of osteoclasts, but after these cells have removed old bone and transferred its minerals to the blood, they apparently have completed their function and disappear.

When osteoclasts have finished gnawing away at the bone, osteoblasts set to work like a team of bricklayers busily troweling on new mortar to patch a breached wall. Osteoblasts are found on the outer surfaces of bones, just below the periosteum, and in bone cavities. At any one time they are rebuilding material on about 4 percent of the skeleton's surfaces. If the bone is damaged in any way, osteoblasts quickly proliferate to lay down new bone tissue. Moving into the tunnel dug out by the osteoclasts, they form new bone by secreting successive layers of the same cement substance and collagen that united in the original fetal skeleton. Several months pass before osteoblasts finish their masonry. When the new matrix is in place, it quickly hardens from calcium brought from the blood by the osteoblasts.

A third type of bone cell, the osteocyte, lies embedded in the hard matrix of bone and acts as the resident caretaker. The most stable of the three kinds of cells, osteocytes begin their life as osteoblasts. As in the original fetal formation of bone, some osteoblasts become caught in their own secretions, which harden around them. They lose their bone-building power and assume a new identity as osteocytes. Occupying small gaps called lacunae, which are connected by tiny channels (the canaliculi), osteocytes help create a flow of liquid calcium and phosphorus through the channels between the matrix and the blood stream. Rendered immobile and altered in function by isolation, osteocytes live quietly, maintaining bone's vascular stability and the silent rhythm of resorption and remodeling. Outwardly inactive, osteocytes are a vital part of the skeletal system, for if they die, the bone will die too.

As the osteoblasts move quickly to build new bone where osteoclasts have eaten old tissue away, they put down new matrix on the part of the bone that demands it most. Through exercise, work or other actions, some muscles and bones work harder than others. Although bones appear solid and incapable of changing their shape, they can mold themselves to match the mechanical forces placed on them. Changes in the conformation of bone can continue past the age when the skeleton stops growing.

Less dramatic, but just as constant, is the exchange of minerals, especially calcium, between bone and blood. Calcium fulfills a variety of functions throughout the body. The muscles need it for contraction, and blood requires it for clotting. Bones maintain their strength by absorbing newly assimilated calcium from the blood stream. Expectant mothers are often encouraged to drink extra amounts of milk because they must supply all the calcium to the developing skeletons of their babies.

Wolff's Law

To keep this delicate chemistry in balance, bone acts as the body's storehouse of minerals. About 70 percent of bone's weight and half its volume consist of deposited minerals. Ninety-nine percent of calcium and 85 percent of phosphorus in the body are found in the bones. The skeleton also lodges traces of numerous other elements. When all the rebuilding and cargo-carrying functions of blood and bones are in balance, scientists say a condition of homeostasis exists.

In 1892, Julius Wolff, a German orthopedic surgeon, formulated a law of physiology that accounts for the effect of physical stress on bones. Briefly stated, Wolff's law holds that bones are sensitive to physical demands and adjust their shape accordingly. Thus, athletes and manual laborers acquire denser skeletons than most people. From continual practice, pianists have especially thick knuckles, while ballet dancers develop slightly enlarged big toes. South American gauchos, it is said, develop spiral grooves on their upper arm bones from whirling bolas, lariats containing stones or metal balls, above their heads.

It has long been known that the structure of the skeleton determines personal appearance and beauty. In order to achieve an effect more prized than what nature has provided, though, some people deliberately alter portions of the body and, hence, their bones. Some Burmese women wear rows of tight metal bands around their necks to enhance their beauty. With their necks elongated a foot or more, they look undeniably statuesque, but their appearance is achieved at the expense of collarbones and shoulders. The unyielding metal rings push the shoulders down into the torso. Vertebrae that would normally be in the upper back are consequently elevated to the unnaturally extended neck. Sometimes, if a woman has been unfaithful or committed some other offense, her brass neck rings are removed as punishment. Her head then falls around her shoulders because the neck muscles have atrophied from lack of use. Unless she supports her head, she could easily die of suffocation from a blocked windpipe. Gradually the woman's neck muscles will strengthen, but her shoulders will remain permanently lowered.

Space flights have shown that the opposite of Wolff's law is equally true: if bones are not used, they deteriorate. In the Gemini V mission of 1965, American astronauts L. Gordon Cooper and Charles Conrad lost more than 20 percent of the mass in some of their bones. In later space flights astronauts followed special diets and exercise regimens, but the weightless environment still caused some bone loss. Mars may never be explored because scientists and doctors have cautioned that extremely long space flights could severely damage the skeleton. It would take nearly a year for a space ship to reach Mars, by which time the body's bones would have deteriorated to a state of near uselessness. In addition, much of the calcium lost from the bones would have to be excreted through the kidneys, raising the possibility of painful kidney stones for outer-space travelers.

Diminution of bone is evident even among people convalescing from a broken arm or leg. A leg in a cast will lose as much as 30 percent of its bone content in just a few weeks. The muscles will similarly atrophy, although the healthy leg will retain or even increase its previous strength. After recuperation, exercise will help injured bones return to normal.

Sometimes, however, exercise can place undue stress on the skeleton. Athletes are especially susceptible to injuries caused by pushing the skeleton beyond practical limits. When teenagers

A frequenter of the Parisian ballet, nineteenth-century French artist Edgar Degas became absorbed in the delicacy and excitement of dance. Lit by the soft underglow of footlights, this prima ballerina arches her back and extends her arms in an elegant arabesque. Anatomists have learned that dancers' movements — from the swift glissade to the dramatic grand jeté — produce adjustments in the skeleton. The hipbone of a dancer, below, shows a cavity carved from repeated leaping. In the Polish Rider, by Rembrandt, above right, the young equestrian holds his knee against his horse. Long-time riders often get "rider's knee" from such gripping. In response to pressure, a bone spur develops on the upper bone of the knee, far right. The spur protrudes from the smooth curve on the bone's left side. Other forms of occupational stress can lead to similar compensations in the skeleton. Pianists, portrayed below right by Auguste Renoir, acquire thickened knuckles from moving their fingers rapidly and from stretching to reach octave chords a hand's-breadth apart. The hand of a masseuse, below far right, reflects, in its exaggerated joints, the pressure and kneading motion of massage.

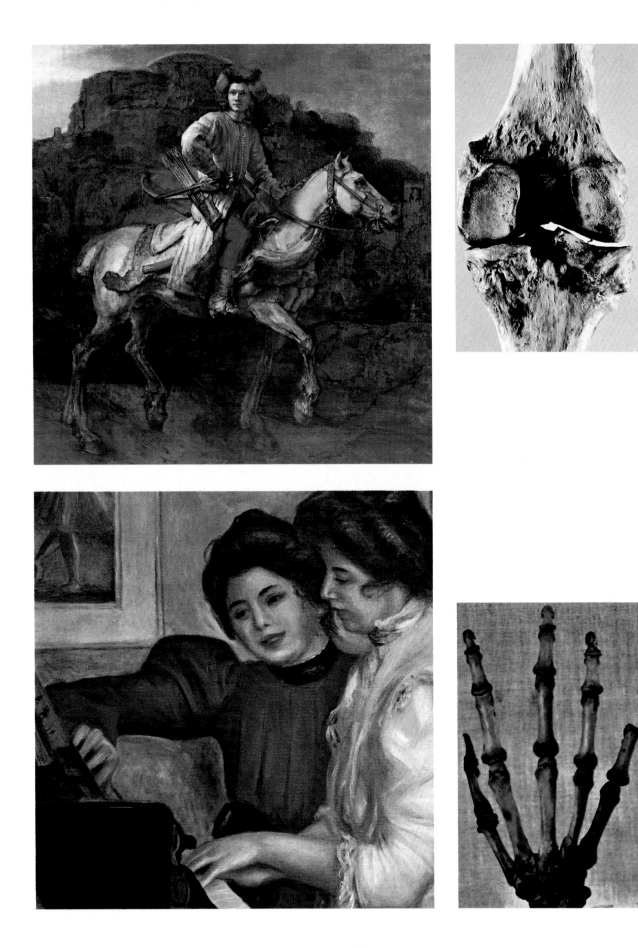

and younger children undergo excessive training with heavy weights, they can damage their bones and, some doctors believe, stunt growth.

The study of stress on bones, both pressure from outside the body and the forces that bones exert on one another, has led to an interesting discovery. When bones grow in response to physical stress, it appears that small piezoelectric currents are transmitted in the bones at points of compression. Electrical charges are thought to stimulate osteoblasts in building new bone to repair fractures. It is possible then that bone may act as a kind of generator, translating mechanical forces into electrical responses.

Whether turned on by a figurative electrical switch or not, osteoblasts are most active early in life, when the skeleton is still growing. Bone formation outstrips resorption, causing the bones to enlarge and assume their adult shape. As the last of the growth cartilage at the epiphyseal plates is supplanted by bone, however, the chemical and metabolic impulses of the skeleton begin to shift.

Up to the age of twenty-one or so, the maturation of bone is so regular that a person's age can be quite accurately judged from the way his skeleton has developed. After adolescence, the feverish proliferation of bone cells slackens and the mineralized matrix of bone starts to assume the chemical stability of adult tissue. As bone growth halts, resorption increases. Mineral content in the skeleton gradually decreases, causing the bones to repair themselves more slowly. Given enough time, too little exercise and a careless diet, the skeleton can become significantly weakened.

Deteriorating Bone

Although bones do not show marked degeneration until after the age of forty, 10 to 15 percent of Americans already have lower bone density by the time they are twenty-five. Some bone loss is unavoidable because of resorption, but doctors say one of the most serious causes is a lack of calcium in the diet. Every day, most American adults lose up to thirty milligrams of calcium, and often more. This loss must be made up by the withdrawal of additional calcium stored in the bones. With too little calcium in the diet, the decaying bone can never be recovered. Because

women have lighter skeletons, they are especially hard hit by bone deterioration. After menopause, their total bone mass drops about 1 percent each year, resulting over time in thinner bones that sag from supporting the weight of the body.

This chronic and widespread condition is known as osteoporosis, by far the most common bone disease. It did not become a serious medical problem until the middle of this century because in earlier times the average life span was too short for osteoporosis to be a large-scale menace. It affects the bones in much the same way malnutrition would. Besides lack of calcium in the diet, cigarette smoking and insufficient exercise can aggravate osteoporosis. In the United States, it afflicts one in four elderly women and one in eight elderly men. The highest incidence occurs among white and Oriental women. Weak bones are more likely to fracture, and nearly 200,000 Americans each year suffer broken hips because of the debilitating effects of osteoporosis. All too often, these injuries lead to other serious complications, sometimes resulting in death.

The danger of this persistent weakening of the bones increases in postmenopausal women because of lower amounts of estrogen, the female sex hormone, in the body. In studies of women stricken with osteoporosis, scientists have determined that estrogen holds in balance a hormone secreted by the parathyroid glands in the neck. When estrogen is absent or diminished, the parathyroid hormone (PTH) has a stronger effect on bones, causing them to lose calcium at a faster rate. Other experiments have shown that doses of sodium fluoride can also be effective in restoring the strength of weak bones.

The most promising treatment for preventing osteoporosis could turn out to be the simplest. Scientists believe that a proper diet, one rich in calcium, might prevent osteoporosis in many cases or at least lessen its effect. Contrary to popular opinion, a number of researchers have concluded that adults actually need more calcium than children to preserve the strength and density of their bones. The daily recommended amount of calcium for adults is 800 milligrams, but most women over forty-five consume only about 450 milligrams a day.

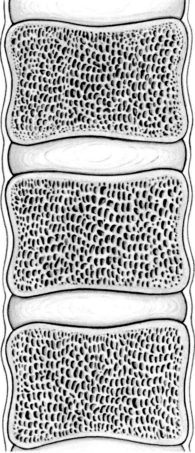

Osteoporosis is the most common disease of the bones, causing them to degenerate and become susceptible to fracture. In a cross section of a vertebral column, left, osteoporosis has so weakened the vertebrae that the cushioning disks of cartilage, seen here as purple, have broken through the thin, decaying bone tissue. Normal vertebrae, illustrated above, are densely packed with a spongelike network of tough bone fibers. The skeleton naturally weakens with age, but adequate amounts of calcium in the diet can forestall the slow deterioration of osteoporosis.

Some nutritionists say that many popular foods — meats, fish, poultry, soda drinks and processed foods — restrict the body's ability to assimilate calcium. The current standard of 800 milligrams per day is probably enough for men and for women before menopause. But post-menopausal women may need to increase their daily calcium intake to 1,200 or 1,500 milligrams to counteract the effects of bone resorption. This amount of calcium is found in a little more than one quart of milk. Other foods high in calcium include yogurt, cheese, leafy green vegetables, sardines and canned salmon containing bones.

The Healthy Skeleton

Since bone must be continually replenished, nutrition is essential in maintaining the health of the skeleton. If consumption of vitamins and other nutrients is too low, bone growth and re-modeling are hindered. The vitamin most important to the skeleton is vitamin D. Ultraviolet rays from the sun convert a form of cholesterol in the skin into vitamin D inside the body. Milk, now usually enriched with vitamin D, is another primary source, especially in winter when the body is exposed to less sunlight. Before milk was fortified, parents bolstered their children's diet with occasional doses of unpleasant-tasting cod liver oil, rich in this essential vitamin. Vitamin D increases the rate of mineral absorption from the intestines and thus increases the amount of phosphorus and calcium in the blood. It also appears to stimulate the complementary action of parathyroid hormone. Although vitamin D causes calcium to be deposited, PTH induces bone resorption. By this carefully regulated antagonism, bone remains constantly renewed.

Lack of vitamin D softens the bones and saps their strength. Without the push that it provides, calcium and other minerals lie idle. Rickets, a disease deforming bones, can ensue. This disorder prevents the proper calcification of bones, causing them to grow soft. Legs characteristically twist outward or bend toward each other, and other bones are similarly weakened. English doctor Francis Glisson first described rickets in 1650, but it was not until the 1920s that scientists discovered its nutritional cause.

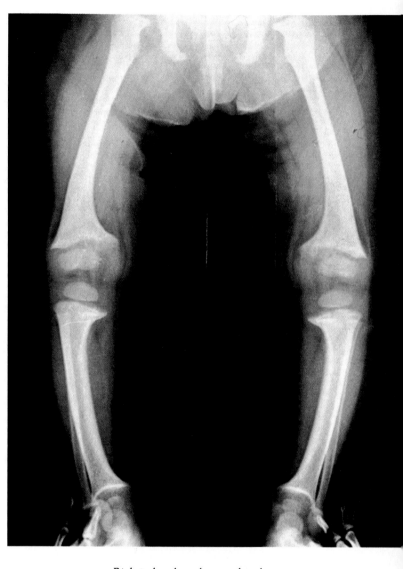

*Rickets has long been a dread
disease, attacking the pliant bones of
children. Caused by insufficient
amounts of vitamin D, it prevents
the bones from hardening normally.
The weak bones of a ten-month-old
child afflicted with rickets, seen in
this X-ray, have bent under stress.*

Rickets occurs only in children. Because the epiphyseal cartilage that causes bones to grow continues to be produced, the spongy, unformed bones flex unnaturally, accentuated by added growth. Without vitamin D, the growth cartilage remains soft, and immature bones, unable to harden, bend under their burden. In adults a similar condition, osteomalacia, can occur from a prolonged deficiency of vitamin D. In rare instances, an overabundance of vitamin D can be present. It can then turn against the body, causing soft tissues — blood vessels, kidneys, muscles and tendons — to harden.

Vitamins A and C also contribute to the health of the skeleton, but to a lesser degree than vitamin D and certain hormones. Vitamin A assists in the development of bones and cartilage, and vitamin C adds to bones' durability by helping to form collagen and connective tissue.

Vitamins often work together with hormones in a chemical choreography to produce the growth and maintenance of bones. Parathyroid hormone and vitamin D regulate bone remodeling, and some vitamin D actually converts to a hormone on its own to promote calcification. Calcitonin, a hormone secreted by the thyroid gland, also helps control the concentration of calcium in the blood, having the opposite effect of the resorption-inducing parathyroid hormone. By inhibiting bone loss and stimulating bone formation, calcitonin is an important component in maintaining the skeleton's mineral reserve.

Other glands also affect the skeleton through the hormones they secrete. Thyroxin, from the thyroid gland, controls metabolism in the body. But because it also governs general maturation of the body from youth to adulthood, thyroxin contributes to the final shape of the skeleton. It helps the bones develop to the correct proportions of an adult. Without thyroxin, a skeleton could grow but it would not mature.

One experiment that illustrates the effect of thyroxin involves the axolotl, a salamander that never enters a true adult state. The axolotl, whose name derives from the Aztec word for "water monster," lives in a lake in Mexico. With gills, a fishlike tail and four short legs, it appears to be evolutionarily lost halfway between a frog and a tadpole. It breeds and follows an aquatic existence in a permanent larval state. But when an axolotl swims in a tank containing thyroxin, it loses its gills, develops legs and lungs and assumes a form never found in nature.

Giants and Dwarfs

When teenagers reach puberty, the sex hormones become more active, spurring on growth and maturation. But by signaling the body that it is approaching adulthood, they begin to shut off the formation of epiphyseal cartilage. After a sudden spurt, the skeleton stops growing as the cartilage calcifies. If puberty comes early, particularly in boys, it can stunt growth somewhat.

Near the base of the skull is the pituitary gland, sometimes called the master gland because it not only sends messages to bones and other tissues but controls other glands to some extent as well. The pituitary secretes a growth hormone, without which calcium, phosphorus and other elements could never unite to form bone. If the pituitary secretes too much or too little growth hormone, it can produce a giant or a dwarf. But excessive height does not, in itself, constitute gigantism. A combination of genetic heritage, nutrition and other environmental considerations can produce exceptionally tall individuals who are not true giants. Many basketball players are more than seven feet tall but they are normal in every way. On the other hand, there has never been a known pituitary giant who passed on his unusual height to his offspring. A true giant's size results from a random physical abnormality.

Giants have excited man's curiosity at least since David slew Goliath. In the early eighteenth century a French academician tried to calculate the height of biblical characters, believing they had been a race of giants. By some stretch of the measuring stick and imagination, he concluded that Adam was 123 feet 9 inches tall.

Giants have long been exhibited for public entertainment. Patrick Cotter, an Irish giant who lived from 1760 to 1806, acquired a certain wealth from performing such stunts as producing dwarfs from his coat pockets. In his youth he worked as a mason, plastering ceilings and shingling roofs without benefit of a ladder. He is believed to

Court portraitist Sir Anthony
Vandyke captured the regal elegance
of England's Queen Henrietta
Maria in 1633. At her side stands
Jeffery Hudson, a dwarf whose
adventures were recounted by poets.

have been between seven feet, ten inches and eight feet, one inch tall. Although he was buried, according to his wish, in a lead coffin under twelve feet of solid rock to keep his skeleton from being disturbed, his remains were later exhumed and studied by scientists. They concluded that, in addition to gigantism, Cotter suffered from acromegaly, a pituitary disorder that causes enlarged bones. His skeleton was found to have a protruding jaw and extremely large hands, two characteristics of acromegaly.

Dwarfism, on the other hand, results from abnormally low pituitary secretions that prevent the skeleton from growing to its full extent. Midgets and dwarfs often suffer from the same glandular deficiency, but the conditions are frequently confused. All unusually small people are dwarfs, but those who are fully formed and have normal proportions are midgets. Other dwarfs have deformities or disproportionate bodies.

Dwarfs have often been objects of curiosity or ridicule, but occasionally they distinguish themselves in other ways. Sir Jeffery Hudson, who lived from 1619 to 1682, led a full and adventurous life. As a boy he became a page to the first duke of Buckingham. Once when King Charles I and Queen Henrietta Maria were guests of the duke, Jeffery astonished the company by stepping out of a pie that had been brought to the table. He was later adopted by the queen and named a captain of horse in the Royalist army at the beginning of the British Civil War. When the queen fled for safety in 1644, he accompanied her to France. Once taken captive by privateers as a boy, he was again captured at sea by Barbary pirates and sold as a slave. In 1653, the brother of a British lord insulted Jeffery by appearing at a duel they were to fight armed with a water squirt. Enraged, Jeffery met the man again and shot him dead. Expelled from the exiled British court in France, he returned to England and lived for a time on an official pension. In 1679, though, he was suspected of conspiring in a plot to overthrow the government and was put in prison. He was released, but died soon afterward. Sir Jeffery was well proportioned but was only 18 inches tall at age thirty. He then began to grow, however, reaching a final height of about 3 feet 9 inches.

Not all skeletal aberrations are caused by unusual conditions present at birth. Degenerative diseases can also take their toll. In 1876, Sir James Paget, a renowned British physician and surgeon, first described a disorder that slowly produced bone deformities over time. Known today as Paget's disease, it strikes about 3 percent of the population over the age of forty, and 9 percent over the age of eighty — second only to osteoporosis in frequency. It often occurs after middle age, but sometimes afflicts younger people, and is more prevalent in men than women.

Paget's disease arises when osteoclasts become overly active, eating away bone. In response, the osteoblasts try to build new tissue as rapidly as possible, but their efforts are too hasty. Bone expands without proper architectural definition, lacking its normal structure. The bones soften, frequently bending under weight. The disease often affects the skull, causing it to enlarge.

A number of physicians have concluded that Ludwig van Beethoven, the famous German composer, had Paget's disease. In his early years he had exceptional hearing, but he was almost totally deaf by the age of fifty, in 1820. Doctors now know that 30 to 50 percent of the people whose heads are affected by Paget's disease lose all or part of their hearing. As the bones of the skull thicken, they put pressure on the auditory nerve, slowly rendering it useless.

Although his physical deterioration started in his twenties, earlier than the age at which symptoms usually become apparent, Beethoven had many other classic signs of Paget's disease. He had short legs, big hands, wide shoulders and a large, asymmetrical head with overhanging brows. His bulky bones appeared to become more pronounced as he grew older, and his contemporaries described his awkward gait, upper torso thrust forward. Words from Alexander Pope's *Essay on Man* can be taken almost as an epitaph for Beethoven and his debility:

> The young disease, that must subdue at length,
> Grows with his growth, and strengthens with
> his strength.

Nevertheless, there has seldom been a more determined genius than Beethoven. He composed his Ninth Symphony, the *Missa Solemnis* and his majestically complex late quartets at a time when he could scarcely hear a note played *fortissimo*.

Because bone must combine strength, lightness and flexibility, it is sometimes pushed beyond its point of resilience. It can withstand shocks of all kinds, but like any living tissue, it is sometimes injured. When tension on a bone exceeds ten tons per square inch, it will often fracture. A bone can break in an almost bewildering number of ways. A fracture may be complete or partial, either breaking the bone in two or breaking it only part of the way. In open fractures the broken ends of bone protrude through the skin, but closed fractures do not surface through it. In a comminuted fracture, the bone is splintered or crushed, the fragments lying between the two larger broken ends. A greenstick fracture occurs only in the elastic bones of children. It is so named because, like a young twig, the bone is

Sir James Paget

An Eminent Victorian

Born in a provincial English coastal city, Sir James Paget rose through hard work to become the most noted surgeon, physiologist and medical teacher in nineteenth-century Britain. He became the friend of royalty, politicians and poets and was surgeon and adviser to Queen Victoria for more than forty years. He lived a charmed Victorian life filled with personal happiness and prominence in his profession.

Too poor to attend a university, Paget was apprenticed at the age of sixteen "to learn the art and mystery of a Surgeon and Apothecary" from a practitioner in his home town of Yarmouth. During his apprenticeship he published, with one of his six brothers, a book describing the plants and insects around Yarmouth. In 1834, when he was twenty, he became a medical student at St. Bartholomew's Hospital in London. He placed first in most of his classes. As a student, he identified trichinae, parasitic worms transmitted through poorly cooked pork, but a senior surgeon claimed credit for Paget's discovery.

Continually short of money, he took on a miscellany of journeyman work: writing articles for reference books, editing medical journals, acquiring cadavers for dissection

and supervising the hospital museum. He taught anatomy and physiology, holding at one time the imposing title of "Demonstrator of Morbid Anatomy." He learned French, German, Italian and Dutch in order to translate scientific papers. In 1852, he entered a private surgical practice that brought him wealth to match his reputation as a teacher.

Throughout his eighty-five years, Paget was an industrious worker, always reading or writing his medical studies at home in the evening while guests and family talked and played music. "I have never once had to leave my family or any quiet party of friends in order to work alone and undisturbed," he remarked. In an age of orators, he was considered one of England's best public speakers. British Prime Minister William Gladstone

once divided mankind in two categories: the fortunate ones who had heard Paget speak and the unlucky ones who had not. Among Paget's distinguished friends were the Prince of Wales, Robert Browning, Louis Pasteur and Mary Ann Evans, better known by her pen name of George Eliot.

In 1856, Paget first saw a patient who had felt for two years a persistent aching in his legs. Gradually his legs became bowed, he began to lose height and his head grew larger. Paget noted that the inside circumference of a regimental helmet the man wore increased by nearly five inches over a period of thirty years. "In its enlargement," Paget wrote in his careful, crisp style, "the head retained its natural shape and, to the last, looked intellectual, though with some exaggeration." When the man died, his mind was "as clear, patient and calm as ever," according to Paget. But autopsy revealed that the bones of his skull were pocked and had become four times thicker than normal. In 1876, the famed surgeon presented a paper describing his patient's condition, which he called *osteitis deformans*. Since that time, the disorder has been known as Paget's disease of bone.

broken on one side and bent on the other. In spiral fractures the bone is twisted, leaving a spiral separation. Transverse fractures occur at right angles to the long shaft of bone. With an impacted fracture, one end of the bone is driven firmly into the other. If the force of a fracture pushes the broken fragments out of their normal position, they are displaced. If the anatomical alignment is undisturbed, the fracture is nondisplaced.

A final kind of fracture occurs without direct injury as a result of fatigue. Stress fractures, as they are known, are relatively common in athletes who do a considerable amount of running. Pain or aching develops at the point of stress, eventually forcing the athlete to give up his routine until the bone heals. Stress fractures are particularly troublesome to basketball players. Professional star Bill Walton was forced to end his career prematurely when a chronic stress fracture led to a series of other foot injuries.

To Mend Broken Bones

Broken bones heal much more quickly in children than in adults. A bone that mends in four to six weeks in a child requires three to five months to heal in an adult. Because bone remodeling and growth are so rapid in children, fractures often repair themselves so flawlessly that X-rays cannot reveal where the bone had been broken.

When a fracture pushes part of a bone far from its normal position, the bone sometimes heals in a condition that doctors call malunion. With malunion an angular bone forms instead of a straight one, resulting in a deformity sometimes noticeable beneath the skin. If the discomfort and deformity are minor, doctors will often leave imperfectly mended fractures untouched. In other cases, physicians can break the bone again, watching it closely to see that it heals in the proper alignment. Malunion occurs most often in adults, particularly when fractures are not treated promptly. With increasing age, osteoblasts form new tissue more slowly and will sometimes cling to the closest bone surface, even though it may be out of line.

Doctors have successfully used bone grafts to heal fractures when part of the broken bone dies or will not form a union with the other segment.

If fresh bone tissue is implanted, it often stimulates the bone cells to remove dead fragments and to deposit new matrix material.

The healing of a fracture is something of a marvel. Old bone fragments are cleared away and new tissue is deposited, giving the broken bone its former strength and alignment. From the moment a bone breaks, it begins to heal itself. Bleeding from blood vessels in the bone forms a clot, or hematoma, within six to eight hours of the injury. The breakdown of proteins at the site of the wound causes a spindle-shaped cuff of cartilage to form around the fracture. This callus, as it is called, forms a bridge between the two fragments, holding them in place. Within forty-eight hours of the fracture, osteoblasts begin to respond, moving slowly toward the callus. New bone first appears at some distance from the fracture before it spreads over the callus, replacing the cartilage with solid bone matrix. Osteoclasts devour blood and bone fragments lost during the fracture and smooth off the bone's rough edges after it has healed. Broken bones should be kept clean and inactive, but the best thing to do is wait patiently while nature takes its restorative course. Doctors have learned that the debris left at the fracture site — bits of blood, bone and other tissue — is necessary for the healing bone cells to take hold. The callus that forms around a fracture can be considered a natural splint because it keeps the broken bone firmly in place.

Immobility has long been recognized as the safest way to treat fractures. Hippocrates taught his Greek students how to apply splints more than two thousand years ago. A standard way of immobilizing fractures employs a plaster of Paris cast. The cast firmly holds the damaged bones in the correct alignment and protects them from external jolts. For broken legs, even if only in the lower portion, casts should extend from the foot to the hip to keep the bones immobile.

Traction, which keeps bones in place with weights, was first used widely in the eighteenth century when it was discovered that bones healed better when slightly stretched. The tension applied in traction counteracts the tightening of muscles around injured bones and prevents the muscles from shortening.

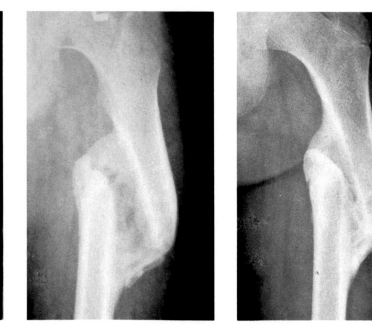

Bones were once treated with methods of almost elaborate crudity. Systems of pulleys, winches, screws and other machines were used to put bones back in place. People were sometimes hung upside down or woven through the rungs of ladders in attempts to set their bones aright.

Hugh Owen Thomas, a Welsh physician practicing in nineteenth-century England, initiated many improvements in orthopedics, the treatment of the skeleton's arrangement and motion. Working before the widespread use of anesthetics, Thomas would unhesitatingly break an improperly healed bone again and set it in its proper place. If he had to correct the alignment of a wrist, he held a wrench behind his back, clamping it on only at the last minute to break the bone in the spot that would make it heal normally. In spite of these rough methods, Thomas devised a number of splints and braces to support the bones and was the first person to recognize that the circulation of blood helped fractures heal by bringing needed nutrients to the injury.

Through the years, the use of X-rays and other medical advances has made the healing of fractures safer and more rapid. Wiring bones together to repair severe fractures gained prominence during World War I. Twenty years after the war it was impossible to tell that soldiers whose faces were shattered in battle had ever been injured.

The skeleton often leaves tales behind in the ways it mends itself. In 1843, African explorer David Livingstone was mauled by a lion, suffering an open fracture of the upper arm. A later injury gave him a "false joint" where his arm had broken. When he died in 1873 and his remains were shipped back to England, doctors identified him from the knotlike formation on the humerus.

Fractures suffered by earlier peoples can sometimes tell us how they lived. Twenty percent of early Anglo-Saxon skeletons show a characteristic fracture in the lower leg. By piecing together several strands of historical evidence, we know that early British farmers twisted their legs when tilling the earth, perhaps to cover seeds with soil. Skeletons from ancient Egypt show that broken forearms, upraised as if to ward off blows, were the most common fractures of that era. The injury is seen most often in women, leaving a sad commentary on everyday life in the sun-washed kingdom of the Nile. The skeletons of the beaten women trying vainly to protect themselves keep history alive in their graves and, like the maids of Shakespeare's *Twelfth Night*, "weave their thread with bones."

Chapter 5

A Measure of Immortality

It is not man's intention to die forgotten. He fights for immortality, struggling to give his voice and vision an enduring presence on the earth. He passes his name on to his children, knowing that although individuals of his kind die, the species continues. He celebrates his survival through the durability of his creations, leaving behind monuments and recorded history designed to speak to the future. Man alone has the chance, as writer William Saroyan put it, to get even with death.

But for all his creative efforts, man's most abiding masterpiece may be the scaffolding beneath his own flesh. Bones are the body's great storytellers. Scattered among stone tools or the debris from some ancient feast, laid to rest in ritual or forgotten by time, bones endure to tell their tales. They are articulate documents that never lack mystery, and often provide the only clues to lives hidden beneath time's shroud.

Relatives of disaster victims wait to claim the remains of their dead. Hunters find skeletons in the woods. Faced with a heap of bones, men ask questions of identity and predicament. Bones tell this and more. They confess mistakes, expose habits, bear witness. Engraved in their surfaces and worn into their responsive forms are stories of many lifetimes. And sometimes, in betraying violent death, they seem to cry out for justice. Bones, as Charles Darwin said, murmur their owner's story "with almost a living tongue." Physical and forensic anthropologists listen well to the whispers. Only in the past century has science begun to ask questions of bones and only recently have anthropologists been able to crack the code of their dusty language.

Most bones quickly succumb to decay. Bleached by the sun, ravaged by scavengers or pummeled by wind and rain, unburied bones soon return to their mute component elements. It is Nature's plan for living matter, and a good one.

"That skull had a tongue in it, and could sing once," grieves Hamlet in the graveyard. "Why may not that be the skull of a lawyer? Where be his quiddities now, his quillets, his cases, his tenures and his tricks?" More than symbols of mortality, skeletal remains, when examined for evidence of identity or history, often provide the only clues to an unrecorded past.

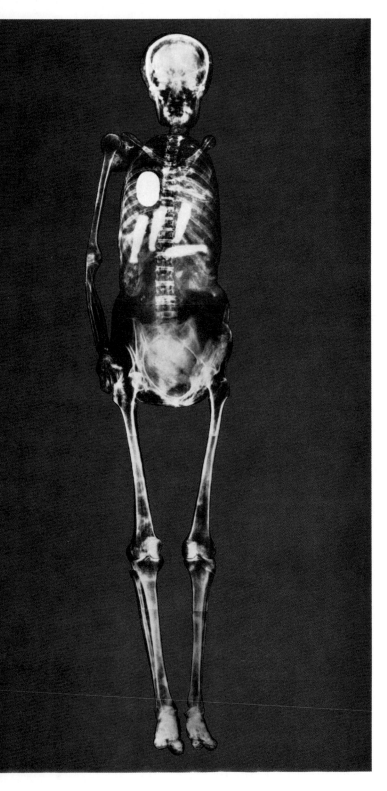

Scientists estimate that the total weight of all living things that have ever inhabited the earth equals the weight of the planet itself. If bones did not decompose, every square foot of dry land would be piled yards high in skeletal remains.

Contesting the Centuries

A few bones, however, survive. There are several types of burial — by chance or careful design — that delay decomposition. The ancient Egyptians were obsessed with preserving a corpse's identity so that its migrant spirits would return to the right body in the next life. They practiced the art of mummification not only on pharaohs but on all the dead. Wondrous monuments, the great pyramids were also designed to protect the corpse from the elements.

But in the earth's laboratory, natural embalming techniques preserve bones longer than any pyramids or potions. While many details of the process of fossilization remain a mystery, scientists do know that the chemical composition of the soil is the most important factor. If the soil is not too acidic, the mineral part of bone will retain its character. Meanwhile, the organic part decays and dissolves through the bone's pores. Seeping in through the same passages, minerals from the soil eventually replace the lost organic matter. These salts then crystallize, cementing the bone from within, sealing it off from further chemical reaction with the soil and preserving its original shape in "stone" that will not significantly change for millions of years. It is little wonder that early fossil finds were often thought to be freak rock formations. The process of mineralization is also one method to date fossil finds. By calculating the concentration of fluorine in the soil, and its concentration in bones, scientists can estimate, to a 50,000-year limit, how long the chemical exchange has been occurring.

In a highly acidic environment, but one deprived of oxygen, the organic part of bone is unable to break down. Mineral components eventually dissolve, but the bone retains its shape. These bones are flexible, rubbery enough to tie in a knot. Soft silt on a lake bottom can often be an ideal acidic bed for preserving bones in this way. If bones are burned, their organic

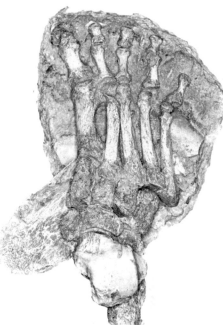

The painstaking work of recovering fossils, left, is often carried out under the most adverse climatic conditions. The finest skeletal remains sometimes favor sites that may be miles from the nearest water, shade or outpost. Above is an example of an especially valuable find, the foot of a Neandertal woman buried 50,000 years ago in southwest France. Anthropologists consider themselves lucky to find one intact Neandertal bone. Unearthing the bones of an entire foot, frozen in their proper places, borders on the miraculous.

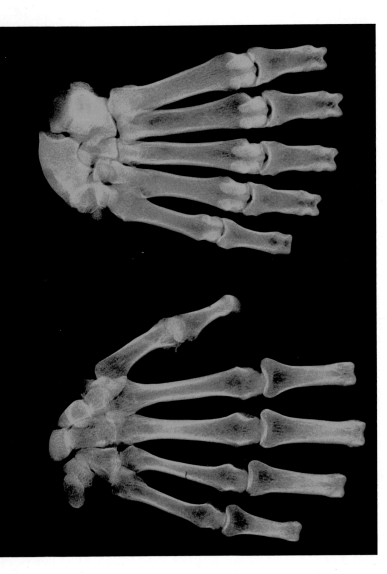

Declawed, skinned and discarded, a bear paw, above, is often mistaken for its human counterpart, top. Telltale differences are apparent in the wrist bones, which are much heavier in the bear, and in the position of the bear's "thumb." Distinguishing between animal and human bones is the first step in interpreting skeletal remains.

components become trapped in place and turn to charcoal. This carbonization actually fortifies the bone, enhancing its resistance to decomposition.

The ideal finds for an anthropologist are remains that have been buried in volcanic ash or gradually and gently packed by successive layers of silt in the sediments of a bog or river. Even though the forces of nature conspire to return man to the dust of anonymity, such coincidences defy the normal course of events and provide scientists with many clues.

Talking Bones

In criminal cases today, forensic anthropologists help law enforcement agencies identify skeletal remains. Most of their techniques were developed for paleontology, the study of fossils and ancient forms of life, and paleoanthropology, the study of ancient man. Despite the availability of various back-up clues, such as dental records and clothing, forensic anthropologists faced with a set of bones prefer to construct a description unclouded by any preconceptions. They let bones alone tell the story. The problems forensic anthropologists address follow a logical order. They first determine if bones are human; then if they are from one person or more; how long it has been since death; if they are recent or archeological evidence; and if the skeleton is complete enough to determine sex, age, "race" and stature. Finally, they look for the cause of death.

Many animal bones look almost human. The feet and hand bones of bears are remarkably similar to their human counterparts. Every fall, hunters appear at sheriffs' offices with evidence of what they believe is foul play. To the untrained eye, limb bones of newborn humans are virtually indistinguishable from those of kittens, puppies or lambs. And the vertebrae, long bones and ribs of older dogs, when found with no corroborating remains nearby, can closely resemble the corresponding bones of a human child.

"What looks human may turn out to be animal," writes Lawrence Angel, "but sometimes it is the other way around." Angel has been curator of the Division of Physical Anthropology at the Smithsonian Institution in Washington, D.C., since 1962, where he has carried out extensive

106

Lawrence Angel

Skeleton Sleuth

J. Lawrence Angel, curator of physical anthropology at the Smithsonian Institution, sees his work as more than just a museum job. When law enforcement agencies bring unidentified skeletal remains to Angel, he feels he is the sole representative of the lost person, especially a murder victim, who has "no way of taking revenge on the murderer, except through identification." He quotes a Verdi requiem to make his point: "Free me, oh Lord, from eternal death." To Angel, nameless skeletons remain eternally dead.

Angel's office lies at the end of a maze of cabinets reaching to the lofty ceilings of the Museum of Natural History. Each cabinet is labeled for sex, age and race and together they contain the largest collection of human bones in the world. The remains of more than 26,000 individuals make up Angel's library.

He is a master of the language of bones. Conjuring from bones the human past or evaluating evidence for the police, he daily adds to their vocabulary. In hand bones alone, he reads a lifetime of wielding a sword, casting a fishing net or operating a factory machine; in leg bones, a career in the saddle; in

a jawbone, the signs of practicing the clarinet. They all tell a story to Angel.

Angel completed a doctorate in physical anthropology at Harvard in 1942, but he traces his introduction to forensic anthropology back to his childhood in England. When he was a boy, he established a small museum. Shells, butterflies, whatever he happened to find around the yard, he displayed and labeled. One day he found "quite a beautiful bone with a socket in it." Identifying it correctly as a hipbone, he put it in his exhibit labeled "cow." When his father, a sculptor, saw the

bone, he told the boy to take it away. It was human.

Today Angel's work spans both past and present. In historical inquiries he relates information garnered from skeletons to the ecologies and geographies of ancient civilizations, combining all available anthropological evidence to show how foods and climates affected societies long dead.

But in criminal investigations, he approaches the evidence cold, avoiding any preconceptions of identity, allowing the bones alone to tell their story. Early in 1981, he worked with artists from the Washington, D.C., police force to reconstruct the features and figure of skeletal remains found in a park. Neither scientist nor artist had seen pictures of any of the suspected victims beforehand. Their portrait matched the description of a tall young black woman who had been missing for two years. They located her dentist. X-rays matched. Identification was positive.

Every year, hundreds of bones under police investigation meet Angel's scrutiny. Cases come to him every week; half are murder victims; only one in ten goes to court. But to Angel the most important part of his work is giving victims back their names.

1 year

2 years

8 years

11 years

anthropological investigations for the Federal Bureau of Investigation (FBI) and other law enforcement agencies. A sheriff once submitted some tiny bones to Angel along with a written confession from a young girl that the bones were those of her stillborn child. Only after careful analysis was he able to conclude that they were the bones of a rabbit.

Several years ago Angel received a package of bones collected from the wilderness near the town of Coeur d'Alene, Idaho. The clothes found with the bones belonged to a Rita M., who had disappeared seven months earlier. Ron M., her husband, was also missing. He was suspected of murdering her. Angel first separated the animal bones from the human — all of which had been gnawed by coyotes. After careful measuring, he assembled the remains of a twenty-year-old woman. But one fibula, the long secondary bone of the lower leg, did not fit. It belonged, rather, to a man about six feet two inches tall. Ron M. stood six feet four. In the trial of a suspect who was later acquitted, it was discovered that most of Ron M.'s height was in the trunk of his body, a fact that earned him the nickname "droopy drawers." The unidentified bone in Angel's reconstruction could have belonged to Ron M., who might have been killed with his wife. Angel could not positively establish an identity from one fibula, but he proved that more than one person was a victim in the case.

Teeth Tell Time

Once it is established that a group of bones are human and belong to one individual, the skeleton is assembled and the four questions of identity explored: age, sex, stature and ethnic group. Aside from obvious growth, the skeleton, throughout its lifetime, goes through various structural changes that occur in a more or less uniform sequence. Deciduous teeth, commonly called baby teeth or milk teeth, erupt and are replaced by permanent teeth. Bones calcify, and by the age of sixty, joints inevitably show signs of arthritis. The plates of the skull meet at seams, gradually fusing together to form a solid dome. An estimate of a skeleton's age is the sum of these telltale signs and several more.

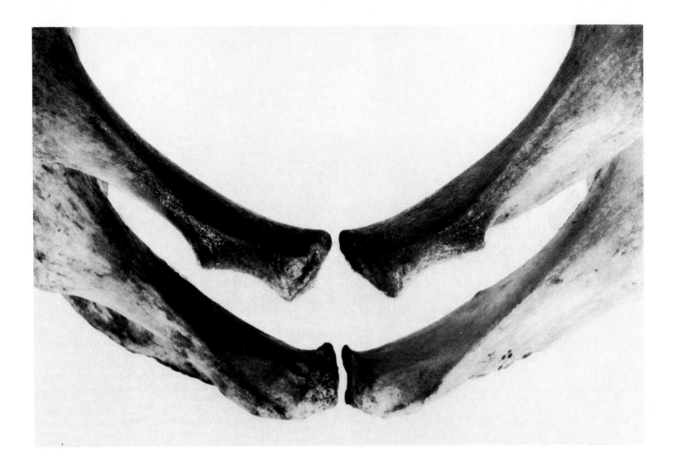

With children and teenagers, teeth yield the most reliable clues to age. The stage of their formation and emergence can even reveal age before birth. When teeth first appear, in the middle months of pregnancy, they are little more than chips embedded in the gums. By the seventh month, the lower front incisors have a shape close to their final form. But by birth, the upper incisors have outpaced their lower counterparts and are almost completely formed. Both top and bottom incisors erupt when the baby is roughly a year old, and by two years, a complete set of milk teeth is usually in place. Between the ages of five and ten, the baby teeth are forced out in a front-to-back sequence, usually with the lower teeth leading the way. Wisdom teeth can emerge any time between the ages of fifteen and twenty-five. Beyond the age of twenty, teeth are more or less in place and assigning age is difficult because of varying rates of decay in adults.

Dental age estimates for children and adolescents have a margin of error of about two years. Furthermore, the teeth of peoples of different ethnic stocks erupt on slightly different timetables, and early influences such as malnutrition can alter the speed of development. But the degree of ossification of the long bones, the femur in the thigh or the humerus in the upper arm, offers corroborative evidence. The shafts of these bones are not fully connected to their caps until late adolescence. Although even before adulthood the shafts and caps of the bones appear smooth, a sign of completed development, the spaces between them are coarse and rough. These areas gradually fill in during adolescence. Like the developmental stage of the teeth, the degree of ossification of the long bones can give rise to a broad range of age estimates, and it takes a trained eye to pick the telling details. Not only do different bones of the body fill in these gaps at different rates, but rates also vary between boys and girls.

Assigning age to adult bones presents an entirely different set of problems. After the age of twenty, skeletal changes occur less dramatically. The massive pelvis, which acts as a center of gravity for the entire body, is actually divided in two parts that meet at a small point called the pubic symphysis. Here, through normal wear and development, changes occur regularly. At the age of twenty-one, just as the age language of teeth grows outdated, deep, ragged furrows crossing the faces of the pelvic bones at the junction point gradually begin to fill in. The two faces are

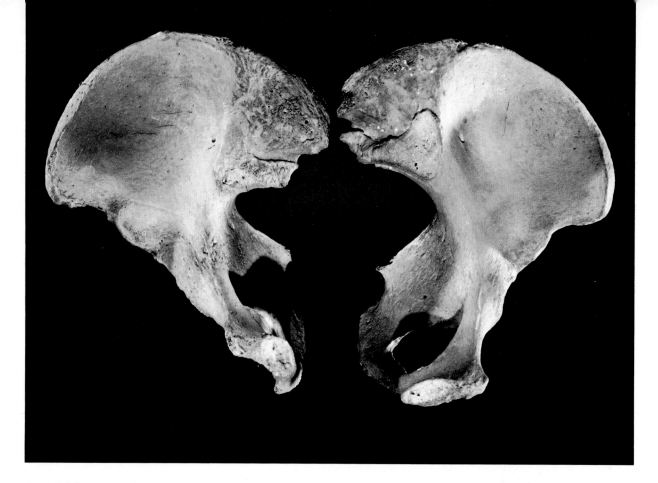

Sexual differences in humans are most pronounced in the pelvis. The mismatched pair of male and female bones above highlights the enlarged and rounded inlet — the birth canal — of the female, right. This evolutionary modification accommodated the development of a larger brain in our species, but illustrates an evolutionary tradeoff. If the female pelvis were any wider, upright walking could be awkward.

smooth by the age of twenty-nine. For the next two-and-a-half decades, rims develop around the edges of each face, reaching completion roughly at the age of fifty-six. From then on, the surfaces of the pubic symphysis deteriorate. Childbirth also leaves notches in these surfaces, engraving on every woman's pelvis a record of the children she has borne.

The skull, for many reasons the most desirable part of the skeleton for a forensic scientist to find, tells very little of age beyond twenty. The brain vault is formed from four major cranial plates joined only by a tough membrane of cartilage during infancy. Throughout life, they grow together to form an **H**-like pattern of seams called "sutures." Although not completely joined, the plates look as though they were sewn together. Over the course of many years, the skull's sutures slowly close from the inside out.

One of the earliest recorded attempts to interpret cranial sutures dates back to the first century when Roman writer Aulus Cornelius Celsus mistakenly concluded that climate had something to do with the formation of sutures. He believed some skulls had none, attributing the lack to warmer climates. In 1865, French anatomist Louis Pierre Gratiolet suggested that intelligence was a factor in the closing of sutures. Incompletely sealed seams signified that the brain remained capable of continued slow growth. Idiots lacked sutures, he reasoned, so "the cranium closes itself on the brain like a prison ... no longer a temple divine ... but a sort of helmet capable of resisting heavy blows."

While there is no definite timetable for the closure of cranial plates, the process follows certain general trends. The coronal suture, the seam running from temple to temple across the top of the head, begins to close roughly at age twenty-four, slows at twenty-nine and stops around thirty-eight. The lambdoid, the coronal's counterpart on the back of the skull, begins closure at twenty-six, slows at thirty-one and is completed by about age forty-two. The sagittal, which connects the coronal and lambdoid like the bar of a capital **H**, begins at roughly twenty-two and ends at about thirty-five.

Male and Female

Nowhere are sexual differences more pronounced than in the pelvis, where they appear even before birth. The sciatic notch — a prominent gap in the rear bottom of the hipbone — widens significantly throughout life, but it widens much faster in female fetuses than in males. Skeletal signs of the birth canal also appear before puberty. In males, the large central opening of the pelvis is heart-shaped; in females it is circular. The pubic symphysis is narrower in women than in men, and women's obturator foramens, the two holes at the base of the pelvis, are smaller and more angular than the rounded ones in the male pelvis.

Sex differences in the skull become apparent only after puberty. Female skulls retain a youthful, graceful aspect throughout life. They are thinner, with more delicate contours, and have a proportionately smaller jaw. Male skulls display more pronounced brow ridges, a more sloping forehead, larger teeth and a larger palate. The male skeleton, in general, is larger and huskier than the female, but this rule has many exceptions. To be certain of sex, forensic anthropologists check other signs dispersed in the skeleton.

Determining height in proportion to other parts of the body is a problem explored by artists

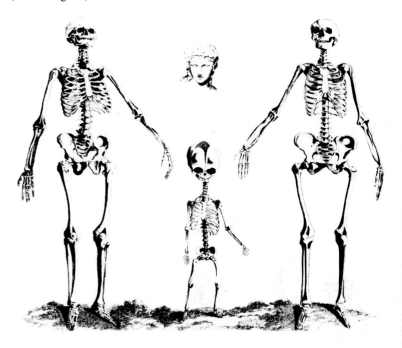

Sexual dimorphism persists in our species after death. A more graceful, less angular skull, a breastbone proportionately broader and shorter and finer wrist bones distinguish the female, right, from the male.

since classical times. Vitruvius, architect of the Roman Emperor Augustus, devised a system that related almost every dimension of the body to every other, a system Leonardo da Vinci believed worthy enough to record in his notebooks: "The whole hand will be the tenth part of the man ... the foot is the seventh part of the man ... from the roots of the hair to the bottom of the chin is the tenth of a man's height. ... The length of a man's outspread arms is equal to his height." These were proportions shown true on canvas and fresco but not exact enough for modern science. More recent attempts have been made to correlate the length of finger, foot and cubit (from the tip of the elbow to the tip of the middle finger) to height. One set of formulas, devised by an anonymous Scottish anatomist at the turn of the century, was based on measurements of 3,000 English, Welsh, Irish "and a few Scots" criminals. His system had a margin of error of about an inch in either direction. A person with a four-and-a-half-inch-long middle finger, by the Scottish anatomist's formulas, should have stood about five feet six inches tall.

A variety of sophisticated algebraic formulas make it possible for today's forensic anthropologists to estimate the height of an individual from

111

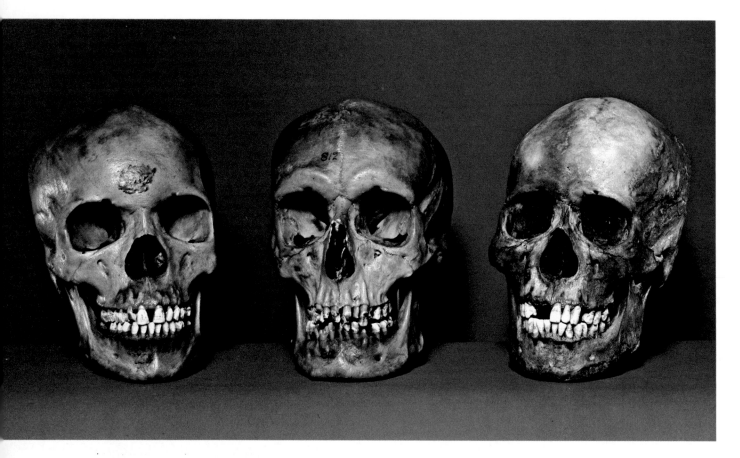

Subtle skull differences between the various human stocks help anthropologists assign ethnic origins to unidentified skeletal remains. Blacks, left, can be distinguished by a well-rounded nasal opening, which comes to a sharp point in Mongoloids, right, and is narrower in Caucasians, middle. The broad, heavy cheekbones of the Mongoloid are another clue, as are Caucasians' smaller teeth. Such traits have a broad range of variation and often overlap within the human family.

any of several bones, but the long bones are the surest guides to accurate estimates of the living height of a skeleton. Presented with a femur seventeen inches long and a tibia fourteen inches long from the same man, an anthropologist could use the formulas to estimate the height of the bones' owner. The femur indicates that the man stood just under sixty-six inches tall; the tibia says slightly over sixty-six inches. Averaging the results, the anthropologist would arrive at a stature of five feet six inches.

With age, sex and stature estimated, investigators turn to the most subtle distinctions of the human skeleton. How accurately can they assign ethnic origins to a set of bones? Here, the skull is the most reliable spokesman. The skull of man demonstrates many variations. At first glance, some differences appear to be adaptive — the pronounced brow ridges of native Australians and Melanesians seem ideal for enhancing the

112

eyebrows' efforts to shade the eyes from the harsh overhead light of the Australian outback or the Fiji Islands. Stronger jaws and the bone structures necessary to support heavier jaw muscles would seem to go along with coarser diets. But such connections cannot be proved.

Today, computers are programmed with more than fifty three-dimensional measurements of a skull. With these proportions, a reasonable guess at ethnic origins can be calculated. University of Pennsylvania anthropologist W. M. Krogman provides a table of cranial and facial traits — guidelines for deciphering ethnic identity — for the major human stocks. A long skull length characterizes Mediterranean, Negroid, Nordic or Mongoloid origins; shorter lengths suggest central European or Alpine peoples. The Nordic, Mediterranean and Negroid peoples also have narrow faces, whereas those of Alpine, and especially Mongoloid, populations are noticeably wider. Negroid and Alpine peoples have wide nasal openings in the skull; those of Nordic, Mediterranean and Mongoloid ethnic stocks are narrower. But, again, within each of these stocks there are further variations, making identification difficult if not impossible for all but an expert.

Skull characteristics suggest some clues to the migrations of ancient man. Cultural anthropologists seek to follow his footsteps by tracing myths, artifacts, laws or any recorded traditions of a people. By examining, classifying and dating human bones, paleoanthropology picks up where these cultural clues leave off. Among all the chores of the paleoanthropologist, the problem of dating leads to the most controversy.

The most commonly used technique of dating a fossil is the carbon 14 method. This special form of the carbon atom exists in all living matter. When life ceases, the living organism stops accumulating carbon 14 and the atoms retained begin to degenerate into nitrogen 14. It takes 50,000 years for all of the carbon 14 in a specimen to completely decay. The amount of carbon 14 decay charts a bone's age on a 50,000-year-long calendar. This seems a very long time, but it is insufficient for the study of early man. A newer system traces the decay of potassium into the gas argon, a much slower chemical process, which

113

Neandertal man has long suffered from the stereotyped image of a slouched, bull-necked lummox, as represented above in Chicago's Field Museum. The Alley Oop *image derives from a famous but inaccurate reconstruction by French anthropologist Marcellin Boule, who used the most complete Neandertal skeleton available at the time, left. The bones were, however, severely bent from arthritis, which led to the misrepresentation of an entire people.*

extends anthropologists' dating abilities back millions of years. Potassium-argon dating was the method used to estimate the age of some of the craters on the moon. Another technique counts the scars that decaying uranium atoms etch in zircon crystals found in the soil surrounding a bone. Today, more than one system is used to establish the most accurate date possible.

Before these sophisticated systems were developed, anthropologists depended almost entirely on geologists to tell them a specimen's age, reached by studying the layers of earth in which the bones were found to determine the geological era in which their owners lived. Without precise methods of estimating age, however, skeletal finds often generated wildly differing opinions among scientists.

The Cave Man

In 1856, quarrymen were sloshing mud out of a limestone cave high on a cliff in Germany's Neander valley, near the juncture of the Düssel and Rhine rivers. The cave entrance was thirty yards above the ground and reportedly small and difficult to enter. The light was probably poor. The workers routinely came across bones, which they dumped aside. But when one particular bone appeared in the mud, workers collected it, along with others lying nearby, and gave them to a schoolteacher who lived a few miles away. The bone was the skull cap of a human; that was certain. But because it was found in a cave, rather than a stratigraphic layer of the earth, it could not be assigned an age by the only dating methods available. The man who had owned these bones came to be called Neandertal for the region in which he was found.

The fascination of Neandertal man lay in his impressive physical appearance — like nothing modern man had seen before. The long bones were unusually massive, with almost exaggerated muscle attachments indicating a tremendously powerful individual. He was bull-necked, slouched over and his knees likely knocked together with each stride. The most remarkable feature was his massive brow ridge. Thus was Neandertal man introduced to the world of the mid-nineteenth century.

115

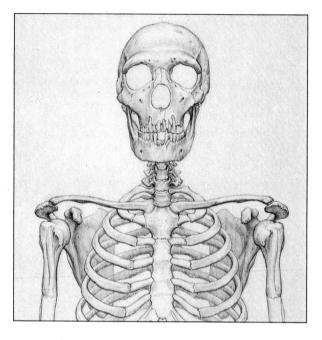

Newspapers began soliciting theories, but since few plaster casts were made of the bones, and access to them was severely limited, imagination led the way. One publication suggested that the bones belonged to a poor idiotic hermit with rickets who died in the cave. Another newspaper characterized him as a most savage and barbarous brute who must have terrorized even the armies of ancient Rome. One of the leaders in the field of comparative anatomy, Professor Mayer of Bonn University, who had not only studied skull similarities between Mongolians and Caucasians but had also enjoyed the privilege of examining the original Neandertal remains, tried to inject some common sense into the public debate. The bones had belonged to a Cossack soldier, he said, a deserter under the command of Tchernitcheff, whose army had been camped in the area in the winter of 1814. The leg bones were bowed not from rickets but from a career in the saddle. The poor soul had obviously been wounded and had crawled into the cave to die, Mayer declared. Tempering his dismissal of Mayer's explanation with good-natured humor, British biologist Thomas Huxley asked why a wounded or dying man would climb thirty yards of steep cliff to die, and if he did, why would he remove all of his clothes and equipment to do it?

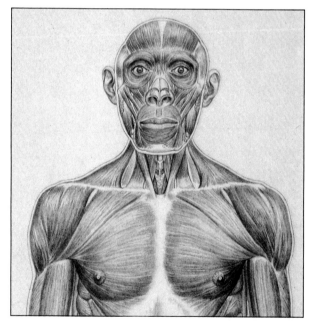

Other investigators soon unearthed skeletal remains of Neandertal man across much of western Europe, from as far north as Belgium to as far south as Gibraltar. Renowned French paleontologist Marcellin Boule reconstructed a fanciful model of Neandertal man based on a set of bones found in France in 1908, by far the most complete Neandertal specimen then available. Fifty years later, pathologists determined that the ancient individual had suffered from severe arthritis. The stooped lumbering posture Boule attributed to this one diseased individual has unfortunately served as an inaccurate stereotype of primitive man to this day. Neandertal man stood upright, with the same posture and gait of a person living today. He also had a brain capacity almost 11 percent greater than modern man's. There is evidence to believe that Neandertal was a spiritual creature. Not only was he the first to bury his dead, but he also cared for the old and crippled.

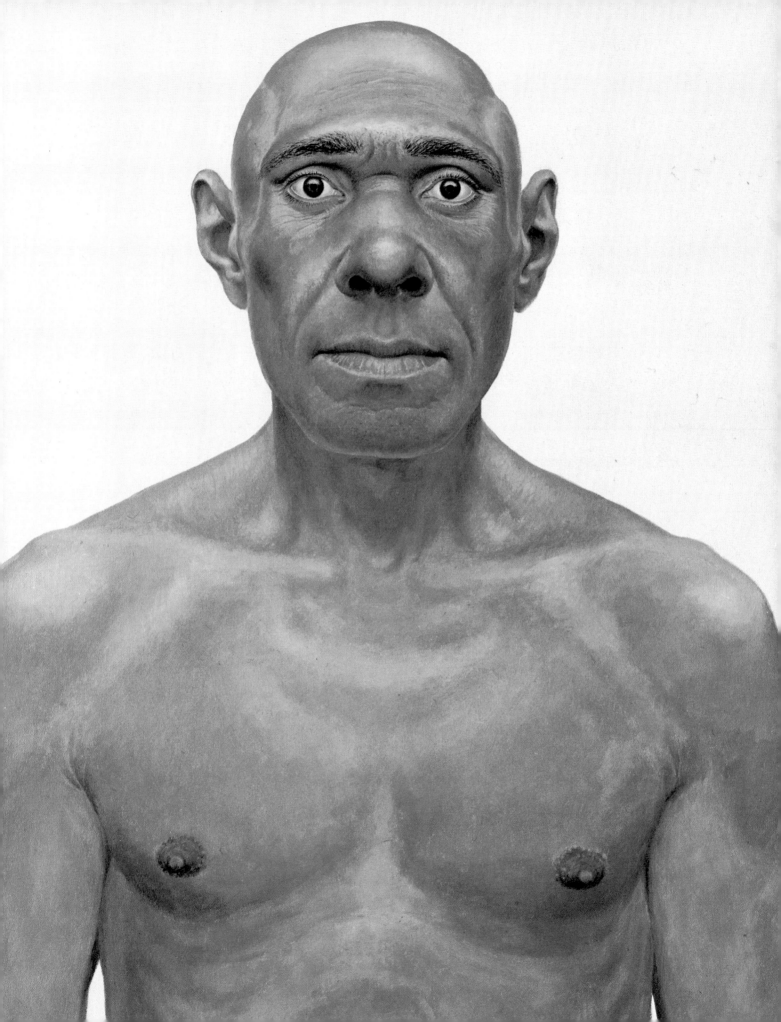

The development of various postures was intimately related to the senses and the environment. One of the first terrestrial quadrupeds was the Ichthyostega, left, who, around 350 million years ago, made its way onto dry land with the help of limbs that were little more than modified fins. Over the ensuing 200 million years, reptiles and early mammals improved upon these clumsy legs, perfecting quadrupedal locomotion. Scientists believe that mammal-like reptiles, such as the eight-foot long Moschops, second from left, were primarily dependent upon their sense of smell, and therefore evolved with a posture that kept their noses near the ground. Some 70 million years ago, the first small mammals took to the trees for defense, represented by the "bush baby," an ancient primate that still flourishes in Africa today. Vision became the most important sense in this new and complex environment; shoulders and joints modified to enable greater mobility; the hand learned to grip; and to coordinate these operations, the brain developed. Increase in size, as with the gorilla, second from right, made tree life more difficult. Longer and stronger arms helped, but when forests began to recede 30 million years ago, some primates were forced to the ground. Arboreal life had moved the center of gravity of some species far enough back to make upright stance and bipedal gait possible. Once upright, hands were free to explore and invent.

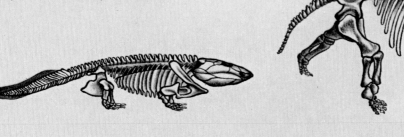

Today, with more than a hundred specimens of Neandertal man to work with, scientists applying the advanced dating techniques know that he flourished on the European continent at least 100,000 years ago. Mysteriously, 35,000 years ago, Neandertal man vanished and was "replaced" (a term anthropologists use cautiously) by Cro-Magnon man, from whom we descended.

Study of Neandertal now focuses on the cause of this sudden extinction. Some paleontologists believe Neandertals were gentle vegetarians who crafted beautiful tools and put flowers on their graves. Other experts argue that Neandertal man's powerful build suggests a vigorous and perhaps a violent life spent in a harsh time. Were Neandertals the victims of mass murder at the hands of Cro-Magnon? Was Neandertal man struck down by plague or absorbed by his distant kinsmen? One of the world's leading paleontologists, Björn Kurtén of the University of Helsinki in Finland, offers his solution in the form of a mystery novel, *Dance of the Tiger*. In his tale of a clash between Cro-Magnon and Neandertal — a battle of species — Kurtén threads together known details relevant to modern anthropological theories of Neandertal man's disappearance, and proposes a solution to the riddle. Hybrid

children of Neandertal and Cro-Magnon parents, according to Kurtén's tale, would have had a special vigor, intelligence and resistance to disease — but they would also have been sterile. Through a combination of biological and cultural chances and mischances, he says, Neandertals would have slowly died out.

Other anthropologists disagree. Arguing that Neandertal simply evolved into Cro-Magnon, they point to early Cro-Magnon skulls, which closely resemble Neandertal skulls but are more graceful — an evolutionary trend that continues to this day.

Every species, including man, has a wide range of variation in any one of its characteristics. That is the beauty of nature and, in this case, the stumbling block of science. Kurtén's speculations, grounded in the best available knowledge of early man, nevertheless point up the main challenge of applying scientific inquiry to the scant remains of prehistoric man. The evidence is vague enough that virtually no informed view can be proved false. One anthropologist likens the search for the story of prehistoric man to reconstructing the story of *War and Peace* from a few pages torn out at random. Such is the case as we go further back into time, as the evidence be-

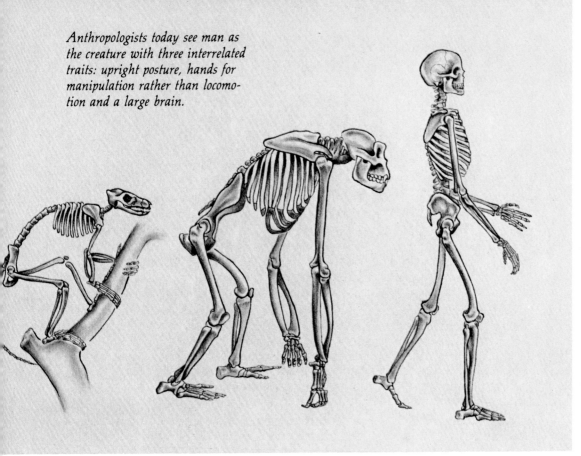

Anthropologists today see man as the creature with three interrelated traits: upright posture, hands for manipulation rather than locomotion and a large brain.

comes more scarce and the question — what are man's origins? — more profound.

Humans belong to the Class Mammalia. We are mammals because we have hair, a heart with two ventricles, four distinct kinds of teeth and our females produce milk. These are traits we share with about 5,000 other species. Because our eyes are directed forward and capable of binocular vision and because we have nails instead of claws, we belong in the Order Primates, a distinction which, again, we share with many other species. And because we walk upright, have large brains, small faces and hands used for manipulation rather than locomotion, we belong to the species *Homo sapiens* — man the Wise — a category we share with none other. To these rather stark taxonomical subdivisions we might add Seneca's definition that "man is a reasoning animal" or perhaps Mark Twain's, that "man is the only animal that blushes. Or needs to." But these are qualities difficult to read from bones.

Three years after the discovery of Neandertal man, an English naturalist published a book that would launch the scientific search for man's beginnings. Charles Darwin's *Origin of Species* was based on tireless scrutiny of the specialization of features in thousands of forms of life. Evolution

Charles Darwin's theories of evolution and the descent of man elicited indignation from philosophers, theologians and scientists alike. His mirror on man also made him a frequent target of satirists.

119

The skull cap of Java man, discovered by Dutch army surgeon Eugène Dubois in 1891, was one of the first major finds in the search for the "missing link." A brain size a third larger than any known ape and evidence of bipedal gait put this half-million-year-old Asian clearly intermediate between man and ape.

was the term Darwin used to describe how living organisms could respond and adapt to stress put on their design by their environment. Nature's method for determining which species and individuals prospered and which declined he termed natural selection, a slow process by which individuals better suited to their local environments increased in number at the expense of less well-adapted individuals. In *Origin of Species,* he seemed more intrigued by the finches of the Galápagos Islands than by man, to whom he made reference only once, writing that "light will be thrown on the origin of man and his history" by the theory of natural selection. The comment was seized upon by overly enthusiastic men of science who, to some extent, misinterpreted Darwin's theories. Convinced that it proved man had evolved from apes in the manner of successive tiles on a kind of primate xylophone, they set off to look for the "missing link" that would definitely tie man to his knuckle-walking cousins.

Asia's Fossil Finds

One of the most remarkable of these quests was fulfilled by Dutch physician Eugène Dubois. In 1890, he took a post as army medical officer in the remote outpost of Java. So obsessed was he with the search for evidence of early ape-men, and so convinced that it could be found in Java, that he suffered the unglamorous post for eight years. He hoarded his free time, spending every available hour studying fossils from sites along the Solo River. Over time, a change in the river's course had exposed a sedimentary level Dubois mistakenly identified as belonging to the late Pliocene epoch, which ended more than two million years ago. The spot was later classified as early Pleistocene, around a million-and-a-half years ago. The layer was rich in mammal fossil remains, including traces of the extinct elephant-like stegodon, the modern-day rhinoceros and Indian elephant, and what he was really after — bones of a manlike creature. Dubois found a brain case and a femur. He telegraphed Holland that he had found the missing link and followed soon after with the evidence.

Dubois's finds were too scanty to create the scientific stir he had hoped for. The skull cap

looked primitive, but the thigh bone was almost exactly like modern man's. If the bones had all belonged to the same individual, then Dubois had certainly found the missing link. But few people were convinced that Dubois's discoveries were the bones of one man. He was honored for his accomplishments, but none would defend him in his claim to have found the missing link. Embittered and dejected, Dubois hid the fossils in his home and let few colleagues examine them. Even when further evidence from the Solo River, including stone tools and ten more skulls, promised to shed new light on Java man, Dubois ignored them. Nor did he respond to another remarkable find outside Peking, China, in the 1920s. Fossil hunters found eight individuals contemporary with Java man, more tools and evidence of fire making and cooperative hunting. Despite the discovery of ancient man in Peking, Dubois charged others with hindering the science of paleoanthropology by proposing new interpretations and operating on preconceptions. But the later finds from Java and Peking were almost certainly related to Dubois's sequestered specimen, and they were clearly man, not ape.

The search for a manlike ape or an apelike man continued. Efforts all over the world were pro-ducing impressive finds, but none had yielded the magic combination. The Piltdown forgery, a case of political sabotage, fraud or perhaps just a practical joke that got out of hand, illustrates the enthusiasm of the chase.

Twenty years after Dubois had uncovered Java man, Arthur Smith Woodward, one of England's most distinguished scientists, and an amateur geologist, Charles Dawson, announced the discovery of an ape-man skull in a gravel pit near Piltdown Common, England. The cranium was human, matched, somewhat incongruously, with a massive jaw. Dawson dated the find about one-and-a-half million years old. Earlier in his career, Smith Woodward had written, "We have looked for a creature with an overgrown brain and an apelike face." Now he had found it. For their discovery, Smith Woodward was knighted, and Dawson, the amateur, was honored by having the new species named after him: *Eoanthropus dawsoni*, Dawson's Dawn man.

As more scientists studied the Piltdown specimen, the feeling grew that the parts were not from the same creature: the jaw, which gave Dawn man his apelike visage, did not seem to belong to the skull at all. It more accurately matched that of a chimpanzee or orangutan (and

121

eventually turned out to be the latter). After years of controversy, fluorine dating revealed that neither skull nor jaw was nearly as old as the hippopotamus or ancient horse remains with which they were found. Closer inspection revealed that a considerable amount of cosmetic work had been done on both pieces; someone had dyed the jaw to match the skull and filed the teeth. Jaw and skull had been chiseled to fit easily together. Dawn man was a fraud, but an expert one. Who did it and why is still a mystery.

For years Charles Dawson was the prime suspect. Recently it has become fashionable to fix suspicion upon one or another of the participants in the original discovery as well as upon some later investigators of the forgery. Piltdown is a cautionary tale of the willingness to believe catered to by those willing to deceive. Well might the deceiver look upon these antics and boast:

> With jawbone of an ass great Samson slew
> A lion, but my deed his feat surpasses,
> For forty years, and with a jawbone too,
> I made our scientific lions asses.

Child of the Desert

In 1924, just as the first whispers of "forgery at Piltdown" were turning into scientific rumbles, an explosion at a quarry in the Taung region of the Kalahari Desert in South Africa uncovered the nearly intact skull of a prehistoric child. Still embedded in limestone, the skull was delivered to Raymond Dart, a professor of anatomy at Witwatersrand University in Johannesburg. He recognized its significance immediately. By incredible coincidence, the skull had rested for thousands of years at just the right angle to fill with eroded soil, bat droppings and minerals, which blended into a natural plaster and created a perfect brain cast. For two months, working with a variety of fine tools, including his wife's sharpened knitting needles, Dart scraped away at the ancient skull to expose its detail.

By the degree of development of its milk teeth, Dart placed the age of the Taung baby — as he named it — at under six years. The teeth were arranged in a semicircular array, a human characteristic, unlike the rectangular configuration of ape teeth. The well-rounded skull had at one

Resurrected from the rubble of a mining explosion in South Africa, the remarkably well-preserved skull of a child, above, gave birth to the species Australopithecus africanus. *Recognizing the human-like teeth and a brain larger than an ape's, Raymond Dart introduced the Taung baby to the world in the early 1920s. "Here, I was certain," he wrote, "was one of the most significant finds ever made in the history of anthropology."*

Louis Leakey

Unearthing Adam's Ancestors

Time in Kenya is measured from dawn: one o'clock in Kikuyu is what Westerners call 7 A.M. Some of the clocks in his parents' house were set to African time, others to European. When the boy rose and dressed to go out into the forests and check the catch of his baited snares, he thought it was just before dawn. So bright was the moonlight on the lush, equatorial highlands that fourteen-year-old Louis Leakey, armed with spear and club, was puzzled not to see any people on their way to market. Hyena snarls from a nearby bush startled the boy into realizing that he had read the wrong clock, it was the middle of the night, and, as he would recall later, he ran home "as though a thousand devils" were chasing him.

Soon after this experience, Leakey lost interest in trapping animals. A book on Stone Age man captivated him and sent him on a lifelong quest that would eventually revise the calendar of man's existence.

While his missionary parents tried to convert the Kikuyus to the Church of England, Leakey was absorbed by the African culture. He went through initiation into warriorhood with his African peer group (an ordeal he would never discuss). Then, following

tribal custom, he built his own mud and grass hut and lived in it alone; he spoke, thought and even dreamed in Kikuyu — a complex Bantu tongue known to only a few white men. Young Leakey was, as the local chief Koinange said, "the black man with a white face." But the two tribal virtues that were to prove most valuable were observation and patience. For his keen eyesight he was given the name *Wakaruigi*, son of the sparrowhawk.

After taking first class honors in anthropology and archaeology at Cambridge in 1926, Leakey returned to East Africa to search for clues to the evolution of man. He was convinced that man's origins lay there. He rejected popular theories that linked all human fossil finds in an unbroken chain culminating in modern man. He believed that our kind has existed in its present form for millions of years, and that other hominid, or manlike, finds represented evolutionary experiments that had failed.

A vigorous, hulking man, Leakey spent most of his adult life on his hands and knees, his eyes close to the ground, looking for evidence to prove his theory. Stubbornly enduring the parching heat, sharing a rancid water hole with two unfriendly rhinoceros, and exhausting their own funds, Louis and his archaeologist wife, Mary, scratched away at the walls of Olduvai gorge with camel's hair brush and dental pick. Thirty years yielded hundreds of stone tools and animal remains, but only inconclusive scraps of what they were after. Their persistence finally paid off. From 1959 on, Olduvai proved a cornucopia of human fossils. Astonishing the world, the Leakeys' finds pushed man's past back to two million years. Leakey died in 1972 at the age of sixty-nine, rightly accorded a preeminent place in the annals of human prehistory.

Prominent drainage route of lakes that once fed the Nile, Olduvai gorge slices through the fossil-rich layers of the Serengetti Plains of East Africa, yielding priceless clues to prehistoric man.

At a time when the first satellites were fueling dreams of man's limitless future, Louis and Mary Leakey's discovery of Zinj, below, pushed man's origins back into the unimaginable past.

time housed a brain with a volume of 390 cubic centimeters, vastly improved over any ape of comparable age. The opening where the spine meets the skull, the foramen magnum, was far enough forward to suggest an upright posture and thus a bipedal gait. The skull, Dart claimed, was 800,000 years old.

"The specimen is of importance," Dart wrote for London's *Daily Chronicle* in 1925, "because it exhibits an extinct race of apes intermediate between anthropoids (i.e., gorillas, chimpanzees, and ourangs) and man." He christened the species represented by baby Taung *Australopithecus africanus* — South African ape.

The attack on Dart that followed, to most of the scientific community, seemed unfairly harsh. His critics charged that he had made too many claims based solely on the skull of a single child. But the objections were soon quieted when a museum curator from Pretoria provided Dart with the corroboration he needed. Robert Broom assembled the fragments of a skull found in another limestone quarry in Sterkfontein outside of Johannesburg. The product was clearly an adult version of the Taung baby. An indefatigable excavator, and a lucky one, Broom continued to search, dig and sift into his eighties. He assembled a rich collection of fossils and established two species of australopithecines, completely vindicating Dart. Taung's relatives were found in a total of three sites. A stockier species, known today as *Australopithecus robustus*, turned up in two nearby locations, Kromdraai and Swartkrans, northeast of Taung.

Leakeys' Luck

Patience, perseverance and thoroughness are all indispensable to a paleoanthropologist, but his greatest asset is luck. One of the luckiest teams ever was that of Louis and Mary Leakey. As the child of missionaries, growing up in the Guikuyu highlands outside Nairobi, Kenya, Louis Leakey was fascinated by prehistoric stone tools. He collected what he thought were flint arrowheads and ax heads. Called "razors of the spirits" by the local Kikuyus, the tools were actually chiseled from volcanic glass. Leakey spent most of his life trying to learn who had made them.

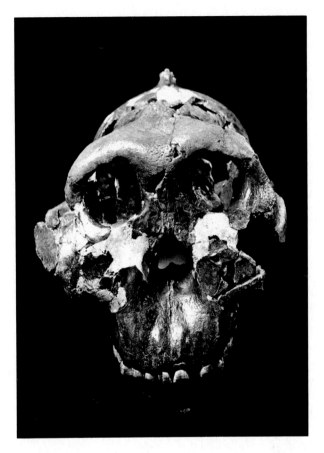

When he completed his studies in archeology at Cambridge in the 1920s, Leakey decided to return to East Africa to search for evidence of early man. With his wife, Mary, he began looking in a gorge in the Serengetti plains of Tanganyika, today Tanzania. Leakey's landscape was an ancient winding canyon which cut through layers deposited by silt and volcanic ash. Like a neatly stacked filing system, the walls of the gorge displayed the Rift Valley's Pleistocene history, the preceding two million years during which Leakey believed human evolution took place. Bristles of bowstring hemp still fringe this gash in the Rift Valley, giving it its Maasai name — Olduvai. Within hours of arriving, Leakey found a large prehistoric ax, but little more followed.

After three decades of "crawling up and down" the walls of Olduvai, "with eyes barely inches from the ground," finding mostly animal remains, the Leakeys' luck changed. One July

Puzzling jigsaw of man's past, finds such as Richard Leakey's 1470, above, form only one piece in the larger puzzle of human evolution. Leakey's wife, Meave, and anatomist Bernard Wood assembled the skull from more than 150 pieces, some the size of a thumbnail. Its owner lived 1.8 million years ago.

day in 1959, while her husband lay in a tent with fever, Mary Leakey spotted a single tooth and carefully resurrected a shattered but remarkably complete human skull from Olduvai's riverbed. With a raked forehead and signs of a massive jaw, it was clearly related to Dart's Taung child. They dubbed their new specimen *Zinjanthropus boisei* (Zinj being an ancient name for East Africa, boisei, in honor of Charles Boise, who had funded the expedition). The Leakeys originally estimated the date of the skull at 600,000 years. Later dating techniques pushed its age back to one-and-three-quarter million years. The horizon of man's past reached back into time unimaginable. Cranial measurements put Zinj, as the Leakeys named him, halfway between man and ape. His head probably sagged forward slightly and he may not have stood as erectly as modern man, but he had better balance than any ape. Zinj had the largest teeth ever discovered in any manlike creature, but unlike those of an ape, the canines were small in comparison to the rest of the teeth. American anthropologist F. Clark Howell hailed the discovery of Zinj as "the event that opened the present modern era of truly scientific study of the evolution of man." *Australopithecus,* the name which now embraces Zinj, Taung and Broom's finds, is believed to have been the first ancestor of man to leave the forests and take up a terrestrial life on the savanna. He averaged four feet in height and probably weighed less than 100 pounds. He likely foraged in groups, using stone tools for defense and eventually fashioning other tools out of stone, wood and bone.

The following years witnessed rapid finds at Olduvai: *Homo erectus,* or Upright man, a relative of Dubois's Java man, and *Homo habilis,* nicknamed Handy man. This man had a larger brain than Zinj and teeth that were more humanlike. There is good evidence that he used tools. He was even older than Zinj.

Ten years after Mary Leakey first spotted Zinj, her son, Richard, exploring a promising fossil site on the banks of Kenya's Lake Turkana, spotted something sticking out of the soil. It was a skull closely resembling that of Zinj. It had taken his parents nineteen days of meticulous digging, sifting and brushing, and eighteen months of patient

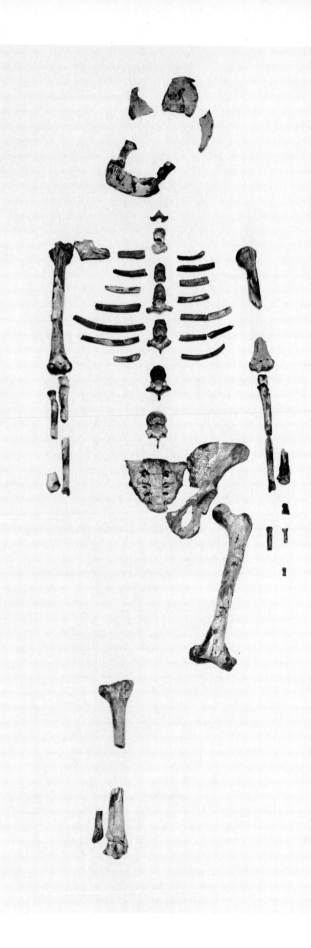

Sifting through the sands of the Afar in southern Ethiopia, members of Donald Johanson's team screen the soil of site 333, which has yielded an unprecedented collection of the fossil remains of man's early ancestors. Here, nearly three-and-a-half million years ago, a large group of manlike creatures met with sudden death. Nearby was Lucy, right, an astonishingly complete specimen. Johanson's team found 40 percent of her ancient skeleton.

128

Donald Johanson uses a dental pick
to chip away the blue matrix
encrusting the cap of a femur taken
from site 333. His finds were to
force a reconsideration of existing
theories of human evolution.

puzzling, to reconstruct Zinj from 400 scraps of bone. Here, at Lake Turkana, Richard Leakey stood holding an almost identical, intact skull. But his find was nestled in deposits 850,000 years older than Zinj. The new skull pushed the age of man back even farther.

A greater discovery was to come years later, in the same area of northwestern Kenya. Richard Leakey set up a camp, Koobi Fora, and sought to attract specialists to the search for man's origins. Today a team of anthropologists, geologists, taxonomists, anatomists, botanists and paleontologists all find work in Leakey's Koobi Fora. The search is on not only to identify early man but to portray him in his environment, to recreate the world in which he lived.

Man's Ancient Image

While Richard Leakey's discovery of Zinj's relative seemed to shed light on the work of his parents, his next find shook up the field as much as Zinj had originally. From 150 fragments — some no larger than a thumbnail — found on a dried slope, anatomist Bernard Wood and Richard's wife, Meave, assembled a human skull. Their recreation was a young, handsome cranium, unlike anything found before. The skull now rests in the Kenya National Museum in Nairobi and is known only by a catalogue number — 1470.

Looking into the face of 1470, we do not see the massive jaws and apelike brow arches of Zinj and Taung — we see our own image. Its forehead vaults are high and a cranial capacity of 800 cubic centimeters, nearly two-thirds the size of our own, puts it far above Zinj (530 cubic centimeters) and the original *Homo habilis* find (650 cubic centimeters) in terms of brain development. The age of the fossil is still a matter of controversy. Richard Leakey claims the skull is almost three million years old, an age that forces a complete reevaluation of how Taung, Zinj and *Homo habilis* fit into man's pedigree. While more evidence is needed to clarify 1470's relation to other fossil remains of early man, Leakey relegated the skull to the ranks of *Homo habilis*. Based on Richard Leakey's estimate of his age, 1470 would have coexisted with *Australopithecus* and could not have descended from him.

About 300 miles northeast of Koobi Fora, in eastern Ethiopia, lies an area called the Afar, known mainly for its sand, rocks and bandits. On a November night in 1974, a little more than two years after the discovery of 1470, an outpost camp in the Afar rocked with excitement. The men were up most of the night, drinking beer and singing along with a tape of a Beatles' song, "Lucy in the Sky with Diamonds." They sang in sheer exuberance, for they had unearthed the bones of an enigmatic female. For the woman in the Beatles' song, they named her Lucy.

The International Afar Research Expedition (IARE), headed by Donald Johanson, Maurice Taieb and Yves Coppens had already had extraordinary luck in the Afar. In three days, they had uncovered remains representing four individuals. They announced that, in light of their new evidence, "All previous theories of the origins of the lineage which leads to modern man must now be totally revised." Richard Leakey countered their claim, demanding further evidence. Lucy, one of several ancient individuals uncovered, was evidence of an extraordinary kind. Not only was she 40 percent intact, she was also nearly three-and-a-half million years old. And perhaps most intriguing of all, there was no

129

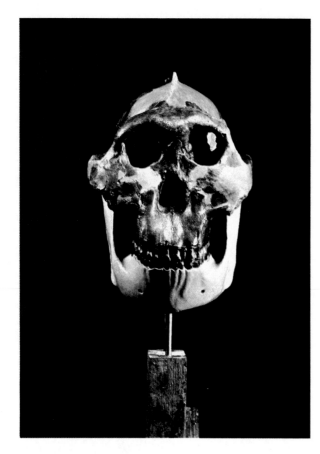

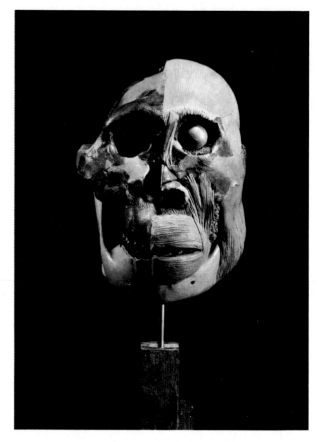

Jay Matternes studied the anatomy of numerous primates to develop the technique he used to flesh out a cast of Louis and Mary Leakey's Australopithecus boisei, *or* Zinj, *above. Many anatomists believe such reconstructions provide our only window on the past. Stalking more evidence of early man, Mary Leakey uncovered the ancient footprints, opposite, in Laetoli, Tanzania in 1978. Geological analysis has dated them at 3.6 million years old, when a mixture of volcanic ash, minerals and water formed this bed of natural cement. Two individuals strode through and a third possibly followed in the footsteps of the larger. The animal tracks off to the side are those of an extinct, three-toed horse.*

niche in any existing theory of human evolution where she could be conveniently put. She was unlike anything that had been found before.

The Family of Man

Just as the chips of each of these skulls were fitted carefully together, and the gaps between them filled in to create a complete portrait, so the entire collection of fossil skulls presents a vast puzzle, stretching over millions of years. Before Lucy was found, there were three popular versions of man's evolution. The first, the single species theory, is the simplest. Taung evolved into Java man and Peking man, both classified as *Homo erectus* and thus qualifying as early humans. Eventually, they evolved into modern man. In this scenario, *Homo habilis,* or Handy man, is grouped with the finds of Dart and Broom, the australopithecines. The only offshoot of the family was *Australopithecus robustus,* who became ex-

tinct about a million years ago. A second theory gave Taung and *robustus* a common ancestor, because *robustus*'s teeth were more primitive than Taung's. The most widely accepted version of man's family tree places Taung as the common ancestor to both *robustus* and *Homo habilis;* the former became extinct and the latter evolved into man the Wise.

Lucy's discoverers contend that she and her kind must represent a common ancestor to both australopithecines and early humans. Johanson and his coworkers call Lucy's kind *Australopithecus afarensis,* and believe that they flourished three to four million years ago. Then their line split, one branch leading to Taung, *robustus* and eventually extinction, and the other leading through *Homo habilis,* Java and Peking man to *Homo sapiens.* Richard Leakey, however, argues that early humans and australopithecines coexisted for more than three million years. They represent two sep-

arate and distinct lineages. He believes we must look back beyond five million years for the common ancestor. One likely candidate for this honor is *Ramapithecus,* an extinct ape that lived in Africa, Asia and Europe twelve million years ago.

Charles Darwin wrote that the only way we could not recognize our parentage — or the only reason we should be ashamed of it — would be if we willfully closed our eyes to it. East Africa has proven to contain the world's richest deposits of early man's fossil remains. Many have called it mankind's cradle. As the sifting and squinting continue, one recent find adds a special poetic twist to the search. At a site along the shores of Lake Turkana, some two million years ago, a small band of man's ancestors made camp in a dried stream bed. It was a temporary home and they left behind many traces, anthropological evidence to the trained eye. Among them is the impression of a fig leaf.

Chapter 6

Form and Future

In his landmark work "Treatise of Man," French philosopher and mathematician René Descartes observed the human body and saw "an earthen machine." Like his "clocks, artificial fountains, mills and similar machines," man's body held not only the intricate gift of motion but the power to intrigue a philosopher's mind. "And I think you will agree," asserted Descartes, "that the present machine could have even more sorts of movements than I have imagined and more ingenuity than I have assigned."

Through the polished prism of modern science, man shares these Cartesian insights. The body expresses a complex interaction of matter and motion, and, in seeking to fathom such forces, we merge mechanics and medicine. Today, surgeons replace joints when they grow diseased and worn. They straighten spines and recraft limbs. They regrow bone and seek somehow to slow its aging. Such bold scientific strokes actually represent little more than the first tentative steps toward man's competent engineering of the human body and toward harnessing the physical ingenuity of the earthen machine.

That the body might indeed be considered a machine is perhaps best expressed in the skeleton's ultimate tendency to mechanically fail. The most prevalent reminder of this is arthritis, a disease with a long legacy. Some natural historians believe arthritis so crippled the dinosaurs as to be the single most important factor in their earthly extinction. Neandertal man, in the mind of the modern observer, walks with a stoop, not because that was his natural bearing, but because the only well-preserved spine of a skeleton from that epoch is gnarled by the disease. The bones of exhumed Egyptian mummies likewise often show the ravages of arthritis.

Arthritis is broadly defined as chronic inflammation of the joints and their subsequent degeneration. Inflammation ordinarily is the body's healthy response to infection — the invasion of toxins, bacteria and viruses. Local tissues in the infected area release chemical messengers which in turn dilate the blood vessels. This makes them more permeable to white blood cells, the body's agents of resistance. The redness and swelling surrounding a wound signals a natural response, the rush of white cells to fight the infection. The persistent swelling around a joint that marks arthritis is anything but.

A Crippling Foe

Less acute than perennial killers such as heart disease and cancer, arthritis tends to take its toll in more silent and subtle ways. More than thirty million Americans suffer from arthritis. Of this number about three-and-a-half million are disabled. The cost of arthritis, though not high in mortality figures, is reflected in twenty-six million lost workdays and $13 billion in medical care and lost wages each year in the United States.

Researchers have identified more than one hundred varieties of arthritis, but for the sake of simplicity it is possible to group them in three general categories: osteoarthritis, rheumatoid arthritis and gout. Gout is the best understood and most treatable of the three. Calling to mind corpulent kings and overfed nobles, gout is traditionally thought to result from the excessive intake of rich food and drink. A rich diet is high in purines, compounds that yield uric acid, which forms crystals should it accrue in the blood in sufficient quantity. The crystals then migrate to joints, where they cause inflammation and pain. For unknown reasons, gout seems to surface most often in the joints of the big toe. Although evidence exists to link a rich diet with gout, the disease is nevertheless an inherited ailment. It does not necessarily tend to occur in overweight people, but rather in those who are unable to rid the blood stream of excess uric acid. Today, drugs that do just that are available, making gout a treatable though not a curable disease.

Virtually every American over the age of sixty shows some signs of osteoarthritis, a disease believed essentially mechanical in nature, in that it results mainly from the normal wear and tear on the joints over the course of a lifetime. This is the most common form of the disease. Of the more than thirty million Americans who suffer significant pain from arthritis, about sixteen million have osteoarthritis, a number likely to increase as people live longer and put cumulatively more stress on their bones.

Osteoarthritis turns up mainly in the load-bearing joints — hips, knees and spine — and in the hands. For some still unexplained reason, it afflicts women twice as often as men. It occurs when cartilage, serving as a shock absorber between bones, begins to break down. Ultimately, bone rubs against bone, causing severe pain and reduced mobility. As a by-product of this process, the regenerative chemistry of bone tissue is mysteriously altered. The edges of the worn cartilage, a soft tissue, harden into bone spurs.

Osteoarthritis is not well understood. It is known to result in part from mechanical stress, but there is no evidence linking occupations that are stressful on the skeleton with incidence of the disease. The opposite may be true. Regular and forceful use of the joints may well enhance their function, thus staving off osteoarthritis. More certain, just as every person has his own set of fingerprints so, too, does he walk through life with a unique stride. The biomechanical interplay of the bones with their own particular skeletal stresses and strains culminates in patterns of skeletal wear that differ from one person to the next. A person with a pair of bones that are slightly misaligned at birth, or who has suffered a severe fracture or injury becomes predisposed to developing osteoarthritis. In this sense, osteoarthritis shares a common root with the often more virulent rheumatoid arthritis. Somewhere deep in the web of genetic factors that give life its diversity lie a set of biological characteristics that create not a direct inheritance of the disease, but a predisposition to it.

In rheumatoid arthritis, the synovial membrane, the tissue lining the joint, becomes inflamed. Malfunctioning white blood cells and joint tissue cells form a deposit called pannus on the joint surface. Enzymes released by the inflammation mysteriously turn on the tissue and digest the joint. Scar tissue can then form between adjoining bones and change to bone,

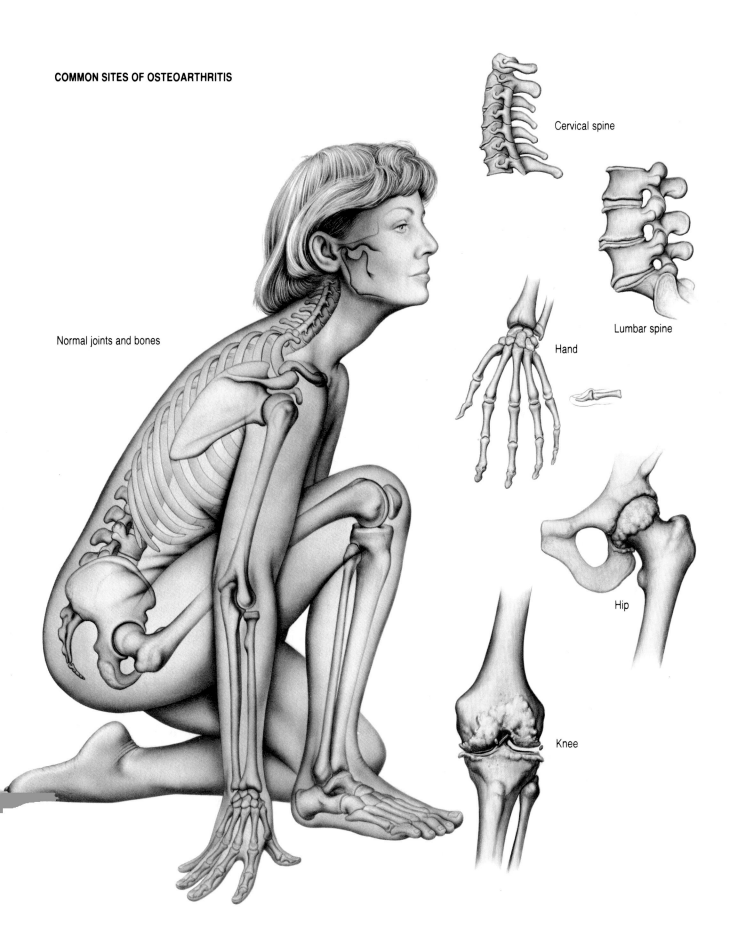

COMMON SITES OF OSTEOARTHRITIS

Cervical spine

Lumbar spine

Normal joints and bones

Hand

Hip

Knee

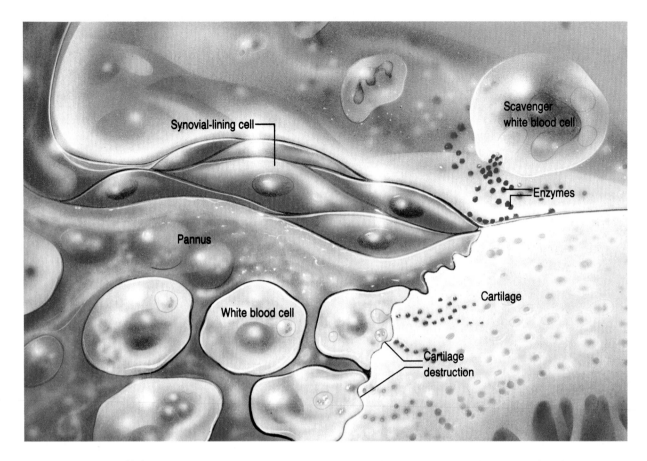

Synovial-lining cell

Scavenger
white blood cell

Enzymes

Pannus

Cartilage

White blood cell

Cartilage
destruction

A deadly deposit called pannus builds in the joint to erode cartilage. A sign of rheumatoid arthritis, it harbors white blood cells, joint-lining cells and other cellular debris. Here, tissue-devouring chemical substances breed. As the cartilage is destroyed, white blood cells, part of the immune response, rush to "scavenge" the remains. But in an arthritic joint they can mistakenly eat healthy cartilage, reinforcing the course of the disease.

which fuses the joint. Rheumatoid arthritis, some researchers currently think, arises from the volatile mix of a yet-to-be identified virus and a faulty immune response. The body's white blood cells are correctly alerted to action by the infection, but they mistake the joint for the enemy and attack it.

The piecing together of the elaborate puzzle that is rheumatoid arthritis underscores the complex interconnections existing at the cellular level within the human body. As understanding of the disease grows, so, too, does the number of leads scientists now follow. Some researchers have identified prostaglandins, hormonelike chemicals that regulate the dilation and contraction of blood vessels, as possible culprits. Arthritis patients seem to produce an abundance of prostaglandins. Other researchers have focused on a single gene named HLA-DRw4, which is known to trigger attacks on collagen, the basic protein in joint and bone tissue. Still others suspect the body's immune B cells, the white blood cells that produce the immune system's antibodies. In a high proportion of arthritics, the B cells have an "activator" that manufactures autoantibodies — immune agents that specifically attack the body. Some scientists have recently isolated a protein produced during pregnancy that temporarily causes remission of the disease in some women.

As the possible factors contributing to rheumatoid arthritis are numerous and often not well understood, so, too, are the remedies. In conjunction with physical therapy, drugs are the most common therapy. Chief among these is aspirin, the most frequently used anti-inflammatory drug with the fewest side effects. Generally, the more a drug reduces inflammation the more pronounced the side effects. More potent and more perilous than aspirin are a range of drugs called corticosteroids, which can swiftly reduce inflammation. But corticosteroids can also produce a host of side effects, including diabetes, thinning of the bones, cataracts and an increased susceptibility to infection. Gold has also proven to be a successful anti-inflammatory remedy, and may even cause the disease to go into remission. A once deadly side effect of gold, the suppression of bone marrow (the ultimate source of the body's immune white blood cells) can now be carefully controlled. Although gold has been administered to arthritics for more than half a century, how the mineral works in combatting the disease is barely understood. It is known that the mineral migrates to the body's joints.

Just as there is no known cause of arthritis there is also no known cure. However, in most cases remedies are available that alleviate the painful symptoms of inflammation and permit the arthritic to live a normal life. That arthritis is epidemic and mysterious in the medical sense are facts best reflected in how money is spent for its study. For every dollar now spent in America on legitimate research of the disease about $25 is spent on questionable or "quack" cures. Copper bracelets, nonprescription drugs, special diets or dry climate — none of these treatments has any scientific basis in helping to cure arthritis. The intensity of the search for a suitable cure is perhaps embodied in a substance called dimethyl sulfoxide (DMSO). An industrial solvent for which an ample black market thrives, DMSO when applied to the skin appears effective in relieving arthritis. But DMSO can also remove paint and dissolve orlon. What it does to human tissue over time is not well understood.

While the sight of a person pained and crippled by arthritis serves as haunting evidence of the scourge of the disease, surgery to relieve such suffering offers more than a glimmer of hope. It embodies the power of science to fight back. In the most extreme cases of arthritis, surgery — specifically the replacement of a diseased joint with an artificial one — is often the only prescription. Currently, about 190,000 joint implants are performed annually in the United States. Most of these operations replace the hip and knee, the major load-bearing joints.

Metals for Motion

The state of joint replacement amounts to no more than the sum of its parts. This is a science that reflects and is governed by the degree of knowledge in a host of seemingly disparate fields: biochemistry, immunology, metallurgy and engineering.

One of the earliest known replacements of a joint occurred in 1891 when German surgeon Theophilus Gluck replaced a diseased hip joint with an ivory ball and socket cemented and screwed into place. The hip summarily popped out under the weight it was forced to bear. In 1893, French surgeon Jules Émile Péan implanted

an artificial shoulder joint made of platinum shafts joined by a hard rubber ball. Two years later he had to reoperate. The tissue around the implant had festered.

The short-lived successes of Gluck and Péan proved hardly atypical for their era. Their operations elaborated two major problems that plagued early joint implantation — undue stress on the joint and the threat of rampant infection spawned by the invasion of materials foreign to the body. In time there followed a progression of trial and error in the use of metals — gold, silver, chromium and copper. The right metal, if it could be found, might withstand the considerable forces generated by skeletal motion and simultaneously not insult the body's biological integrity.

If the modern age in this metallurgical search has a beginning, then perhaps it came in 1938. That year two American surgeons, W. G. Stuck and C. S. Venable, developed an alloy of cobalt, chromium and molybdenum. They gave it the trade name Vitallium and found it to be virtually inert in body fluids, which meant that it did not readily decay, break down or become reabsorbed. It did not trigger a massive immune response so characteristic of many other materials. Vitallium and two other metals — stainless steel and titanium alloys — today serve as the three metal compounds used in artificial joint replacement.

An Elaborate Accident

Along with the emergence and evolution of suitable prosthetic metals came the development of antibiotics, chemical substances that can combat infection and so diminish the risk of surgery. Prime among these was penicillin, an antibiotic as miraculous in its microbial power as in its discovery. The discovery of penicillin was an elaborate accident that occurred in the fall of 1928 when a British bacteriologist, Alexander Fleming, returned to his London laboratory from a vacation. On his windowsill were some laboratory dishes he had failed to empty and clean before leaving for his holiday. They held cultures for study of the bacterium staphylococcus, the most common cause of localized infection in the body. In one dish there lay a strange mold. Fleming didn't know where it had come from, but it was clear that the mold was dissolving the staphylococci germs closest to it. He identified the mold as belonging to the genus Penicillium and named it penicillin.

How had it gotten there? Most likely, he thought, the mold had been blown up two flights of stairs from the laboratory of C. J. La Touche, an Irish scientist experimenting with different molds in the search for a cure for asthma. The landing of the mold in Fleming's laboratory marked the convergence of numerous factors which together combined in an epic discovery. The mold that found its way into the dish happened to be the only species of thousands of Penicillium molds that can make penicillin. The staphylococci germs had not been properly incubated at body temperature — which Fleming routinely did — a procedure that would have rendered them immune to the mold. On the day the mold landed in the dish, the air temperature fell within a vital five-degree-centigrade range that permitted its survival and growth.

Although Fleming seemed aware of the discovery and its potential, the purification and mass production of the substance were beyond his limited resources. He continued to use crude penicillin in his laboratory, but it was not until a decade later, with the impetus of World War II and the support of American mass-production techniques, that the widespread use of the drug became possible. Thus was discovered the substance which stands as a synonym for antibiotic.

Today, there are about 100 different antibiotics available, a dozen of which are most often used. The surgeon operates in a room where the air can be exchanged and purified as often as seventy times an hour. Although infection contracted in the course of surgery remains a serious problem, its incidence has shrunk dramatically. Less than a century ago it was the rule; it is now the exception. In less than 1 percent of all orthopedic operations does infection surface and require the implant's removal.

Major strides in the understanding of orthopedics, the development of sturdy metals and potent antibiotics still would not have assumed such importance had they not evolved in concert with man's mechanical appreciation of the skele-

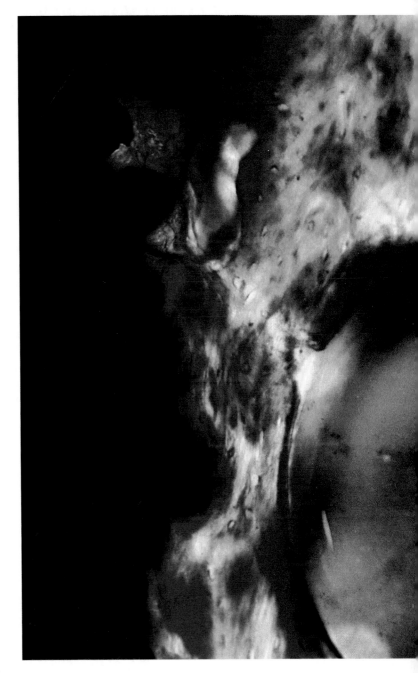

Alexander Fleming

Finding the Magic Bullet

Born the son of a Scottish farmer in 1881, Alexander Fleming grew up in the firm tutorial grasp of nature. The rigors of soil and climate honed his instincts; the turning of the seasons heightened his powers to see. Such an elemental education would, in time, help him discover one of the greatest contributions to medical science, penicillin.

Leaving the land at thirteen for London's higher hopes, Fleming took a job as a shipper's clerk. Soon bored by the work and enriched by an uncle's modest legacy, he enrolled in London's Polytechnic School and from there won a medical scholarship to attend London University.

Joining the staff at St. Mary's Hospital in London in 1906, he worked in the inoculation department where he made vaccines. Infection, then random and rampant, was mankind's dreaded microbial enemy. Science fought back through vaccination, the limited exposure to disease to build immunity against it. A second and more experimental line of bacteriological inquiry was the quest for a "magic bullet," a substance that could be purified as a drug and selectively kill infection without harming healthy tissue. Fleming's superior at St. Mary's and an ardent

believer in vaccines, Almroth Wright, dismissed such a search, warning researchers: "Drugs are a delusion."

Fleming, in time, thought otherwise. He had seen firsthand in World War I how immunity had crumbled under the assault of infection. Nature, he felt, must have other defenses. If man could find them, he could put them to use. Fleming set to work, combining different microbes, watching for that random substance that could kill a host of bacteria. Finally it came — on the wind — a mold floating in through the door, landing in a dish of infectious staphylococci germs and dissolving them. Diluting the mold into gradually weaker concentrations, Fleming found that it still retained its potency. Quietly elated, Fleming nonetheless knew he had to isolate

the mold's bacteria-killing essence to prove his point and silence his doubters. But he was no chemist, and the mold was a mercurial, unstable substance, given to quickly breaking down and changing structure. Over the next ten years, Fleming sought to persuade chemists to tackle the problem. Some were indifferent. Others tried and failed. Then in 1939, Ernest Chain, a chemist who had fled Hitler's Germany, read Fleming's paper on the initial discovery and was intrigued. Working with Howard Florey, an Oxford pathologist, he isolated penicillin as an antibiotic a million times more potent than Fleming's first mold. A Rockefeller Foundation grant of $5,000 and the introduction of freeze-drying, a technique that helped stabilize the mold, greatly eased their work.

The year of their breakthrough was 1940, and Britain feared invasion by Hitler. Florey and Chain soaked their clothing in the mold, should they need to flee. But the invasion never came, and the production of penicillin proceeded. By 1943, the Allies had a drug that would save the lives of countless wounded soldiers. In 1945, Fleming, Chain and Florey shared the Nobel Prize for medicine.

ton. Inherent in the early surgical work of such men as Gluck and Péan was the search to fix the implant firmly in place and to reproduce the nearly frictionless movements of the natural joints. Early in the century, Finnish surgeon K. E. Kallio experimented with skin drawn over the head of an implanted femur to reduce friction in the hip socket. His American counterparts, meanwhile, tried sheets of fat, fibrous tissue called fascia, gold foil, nylon and glass.

In 1960, English surgeons George McKee and John Watson-Farrar, attempting a total replacement of the hip joint, implanted an artificial femoral head and socket. Both were made of stainless steel; but in due time the prosthesis wore out and the screws fixing bone to cement came loose. Then, in 1962, their countryman John Charnley, after three decades of experimentation with different materials, matched a stainless steel femur with a high density polyethylene socket, making a prosthesis that would not wear down rapidly. Six years later, he developed a plastic bonding cement called methyl methacrylate to replace traditional screws, and today it is the most widely used substance for bonding prosthesis to bone.

Charnley's work heralded a new era in orthopedic surgery, one that takes its name from the substance of his work — low friction arthroplasty. Based on the duplication of the strength, mobility and durability of the joint, his technique stands as a qualified success. While implants can now be tailored to fit most joints, they often last no more than a decade before eroding and loosening. The implants restore far more comfort than mobility to the arthritic. The science of biomechanics — the interplay of skeletal matter and motion — is a marriage of forces that are subtle and sheer, long promising to test the limits of man's engineering prowess.

Today, art and science merge in the form of biomechanical engineering. Through a procedure known as gait analysis orthopedists seek to fathom the secrets of skeletal motion. In one such method, subjects with diodes fixed to certain parts of their bodies walk a prescribed path in a laboratory. The diodes, simple electronic devices, send out pulses of infrared light. Cameras sensitive to the light trace the luminous trails the sub-

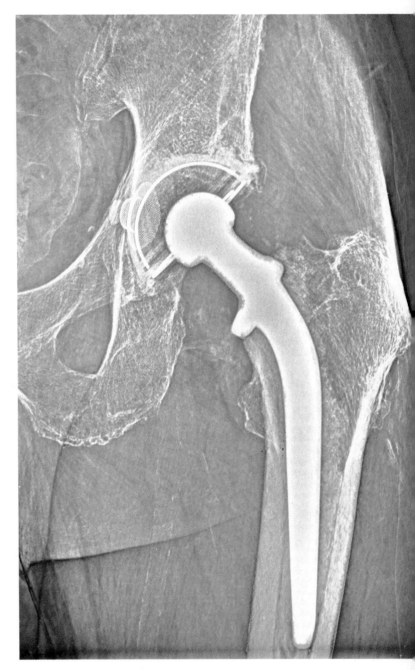

In the stainless steel setting of the Navy's tissue bank in Bethesda, Maryland, a technician checks vacuum-packed, freeze-dried bone, above. From a freezer's misty deep rises one of the Navy's 40,000 specimens of cadaver bone and other connective tissue, opposite page. Freeze-drying and freezing, intricate processes developed over the past 40 years, now ensure bone's successful "banking." Bone can be preserved for grafting and treated to limit an immune response in the patient.

jects make by recording the position of each diode relative to a fixed background — as many as 315 frames per second. The cameras feed the sum of those impressions to a computer, which attempts to mathematically analyze the gait in the form of a three-dimensional figure. To yield such an analysis, says biomechanical engineer Thomas Andriacchi of the Rush-Presbyterian-St. Lukes Medical Center in Chicago, "We've had to make some assumptions. Limb movements are much more complex than our computer model can handle."

Since 1975, Andriacchi has been collecting data through gait analysis on people with diseased or implanted joints. That way he can assess and prescribe the design of an artificial joint. Through his work, Andriacchi sees Nature grudgingly yielding her secrets. He points to the hip which appears as a simple ball and socket, yet with each step as much as six times the body's weight thrusts onto the joint, coming down on a succession of points that often measure no bigger than a dime. In this context, Charnley's first hip implants, although landmark operations, revealed their mechanical limits. They made non-ambulatory people suddenly mobile, but they soon wore out and loosened under the magnitude of forces coming to bear on the joint. Similarly, the first knee joint replacements were designed as simple hinges, but they also failed because the knee moves simultaneously along more than one axis. There are now some 400 artificial knee joints available, pointing to the inescapable conclusion that the intricate design of the joint is something man has yet to satisfactorily match.

Banking Bone

A joint like the knee confronts the surgeon with tradeoffs and dilemmas. Medical wisdom tells him to leave as much tissue as possible when implanting an artificial joint to conserve mobility. But an artificial joint cannot simply swim in a sea of natural tissue. It needs enough of its own kind of matter for anchor and alignment. What is sacrificed in mobility is gained in stability. Likewise, methyl methacrylate is a cement that loosens in a decade or less. It forms a hard, flat bond with bone, like mortar against brick, but does not pen-

etrate and anchor in the tissue. To achieve such an effect, surgeons are currently designing artificial prostheses with recesses sculpted into their surfaces. They hope that bone will grow into these spaces to form a sturdier bond than any cement. But bone takes time to grow. While the implant waits for it to grow and fill those spaces, the joint is rather weak and unstable. And once fully integrated with the bone, the implant is difficult to remove should infection develop.

To surmount such problems of design, some orthopedic surgeons have begun transplanting human bone in larger amounts and more complex ways. Bone has been one of the most frequently transplanted, or grafted, of tissues. It is used to fix congenital defects, patch nonhealing fractures, or fill holes left by the removal of tumors. Ideally, the best grafts are autografts, those in which donor and recipient are the same person. But autografts have one significant drawback. As

the recipient bone rightly recognizes the grafted bone as its own, it often resorbs the graft and leaves a hole unfilled or a break unmended. This then requires a series of grafts with the hope that a little of each will "take," leading to an incremental build-up of new bone. But the amount of bone that can be harvested from a person's body is small due to the threats of excessive scarring and infection, in addition to the severe pain that accompanies the procedure. In young children, because of their relative lack of bone mass, autografts are impossible.

Science has lately sought to circumvent this problem by making allografts, or transplants of bone between members of the same species. This has largely been accomplished through preparation techniques that eliminate the foreign identity of the allograft. Thus it prevents subsequent reabsorption by the recipient bone. Improved methods of bone banking also limit the graft's ability to incite an immune response. The banking of bone is now a commonplace and elaborate process. Banked cadaver bone can be freeze-dried and then stored at room temperature, the end point of a carefully controlled, two-week-long process which sees the sterilization, slow freezing, dehydration and vacuum packing of the tissue. Cadaver bone is also stored through deep-freezing the tissue at temperatures between –70° and –80° C. Ultimately, the preservation of bone by these methods kills off virtually all cell matter in the bone and reduces the potential for it to incite an immune response in the recipient tissue. But collagen, a very primitive tissue, is preserved. This protein is a basic building block of all mammalian tissues and thus readily transferable from one member of a species to another. Collagen is the essential protein that gives bone its structural framework.

Henry Mankin of the Massachusetts General Hospital in Boston is an innovator in the field of large bone grafts. Since 1971, he has performed 133 transplants involving cadaver bone and the replacement of a joint or a segment of a bone. Mankin says bone-banking improvements in the past decade, which better preserve cartilage and decrease the threat of rejection, have increased the success rate of this operation from 50 to 80 percent. What Mankin terms success is a possibility that does not even exist with many artificial implants. With his techniques, the tissue revascularizes; capillaries rethread through the caverns of the bone, giving it life and strength once again. Mankin does not replace joints diseased by arthritis because artificial joints can return to elderly people as much mobility as they need. He concentrates instead on younger patients who might have suffered a bone tumor or a damaged joint. His aim is to replace a joint and restore its full range of motion.

A Protein With Promise

Mankin's work, ambitious and intricate, really serves as a faithful reflection of the miracle of bone. While almost every other tissue in the body heals itself with scar, bone mends itself with bone. Such an adaptation, believes Mankin, is a characteristic essential to man's survival as a species. What sparks regeneration is something known as bone morphogenetic protein (BMP). For close to a century it eluded the probings of scientists because it lay hidden in the densely woven fabric of bone. Then, in 1970, UCLA researcher Marshall Urist and some of his students discovered the protein by chance.

For twenty years, Urist had been trying to induce demineralized bone to remineralize itself by immersing it in an array of chemical solutions. One of his students, a Belgian researcher who had come to California for a year and had fallen in love with the novel setting of the desert, was in the habit of working long hours for two or three days at a time and then taking off several days in a row to wander in the desert. So he conducted his remineralizing experiments quickly and at cold temperatures to guard against contamination by bacteria while he was away. The result, like Fleming's discovery of penicillin, proved a fortuitous convergence of conditions. The cold preserved the BMP while temporarily inactivating the enzymes that degrade it in the bone-banking process. BMP had eluded detection in traditional bone-banking cold storage because after about seventy-two hours the enzymes become active and break down the protein.

Urist is currently at work isolating and purifying BMP, a substance he thinks will revolutionize the surgical reconstruction of bone. The product will be a demineralized powder or chips of demineralized bone that can be used in the operating room as readily as suture material. It can be implanted alone or in the cavities of a prosthetic device to encourage bone growth in the implant and thereby strengthen the bond. The use of demineralized bone is currently undergoing clinical trial. Urist has implanted it in 135 patients having long-bone defects. Of these, he reports healing in 80 percent. Surgeons at Harvard Medical School have implanted demineralized bone in forty-four patients, mostly children with congenital facial flaws, and report bone healing in at least thirty-one cases. They caution, however, that a minimum two-and-a-half-year follow-up period is necessary to judge the efficacy of the treatment.

Just how BMP works remains an unanswered question. Researchers do know, however, that it

Marshall Urist

Bone's Healing Spark

"In science," says Marshall Urist, "the role of chance is overrated. Discoveries are made by prepared minds." For twenty-two years, Urist, director of the bone research laboratory at the University of California at Los Angeles, prepared his experiments, and so his mind. What he sought was the regenerative essence of bone, the means by which this tissue, unlike almost all others, healed with its own cells and not with scar. The secret lay deep in the body's biochemistry, and in 1970, Urist and his coworkers finally found it in a substance they subsequently called bone morphogenetic protein (BMP).

A year out of medical school, Urist found himself far from the quiet cloister of academia. Thrust into the theater of European war as a field surgeon, he "followed Patton's tanks, fixing fractures day and night for four years."

Amid the smoke and rage of battle, the stream of wounded men through his tent, Urist found quiet moments of hope. Penicillin and blood transfusions, newly introduced, gave the battle surgeon life-saving tools he had never before had. "I also learned what bone is capable of," recalls Urist, referring to the tissue's deep-seated healing instincts. "This

is something that no one can watch without enormous admiration for what nature can do in the skeleton's repair."

Admiration, in time, led to devotion. Urist went to UCLA where he spent a decade studying lower vertebrates — animals more successful than man at regenerating skeletal tissues. Then he moved up to human bone. He prepared experiment after experiment, seeking to mimic the conditions that nature creates for bone repair.

Major scientific breakthroughs, says Urist, occur only after research in a field achieves a "critical mass." He credits his associates, graduate students and other researchers with helping him build that critical mass in the study of bone. And he counts his good fortune: "I can think of eight or ten people who have been close" to BMP's discovery. What ultimately misled them were small variations they made in preparing their experiments. Urist cites as one example Nicholas Fenn, who, had he used the tissue of small laboratory animals and not mongrel dogs, would have found BMP. Fenn, like Urist, had been a battle surgeon, an experience that had aroused in him an awe of bone. He, too, mended the wounded, decades earlier during the Spanish-American War of 1898.

The discovery of BMP, Urist believes, signals a new era in the practice of orthopedics. While most applications of basic research in the field have been mechanical, biochemistry now promises to serve as a fresh source of orthopedic insights. Scientists will hone their research to understand bone and heal it at the most elemental level, the cell. Says Urist of man's newfound biochemical knowledge of bone: "There are hundreds of discoveries waiting to be made — by people with prepared minds, and a little bit of luck."

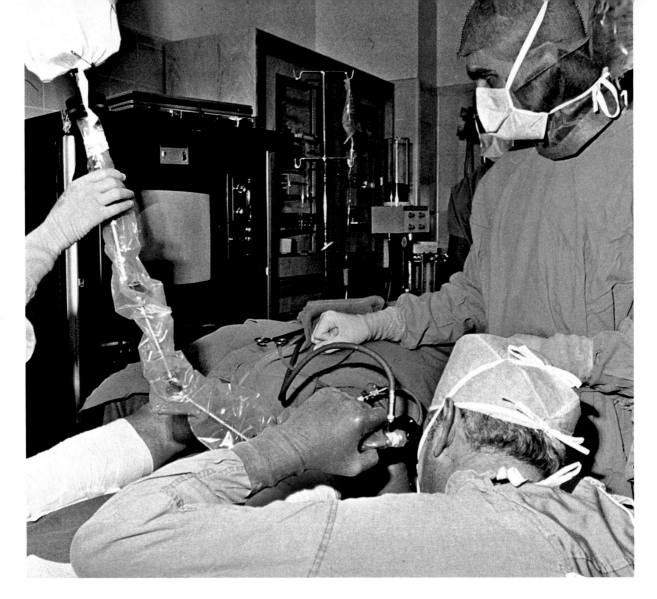

works by the principle of osteoinduction. The protein induces fibroblasts — bone cells that have healed as scar in the wake of a tumor or bad fracture — to begin functioning once again as chondroblasts, normal cartilage-producing cells. This conversion takes about two weeks. In another four to eight weeks, blood vessels begin to pierce the newly formed tissue, completing the conversion to bone. Charles Huggins, University of Chicago oncologist and Nobel laureate, believes BMP is primarily physical, that it triggers a unique electrochemical charge along the surface of the fibroblasts. The charge changes the permeability of the cells' membranes and, hence, their physical expression. Urist disagrees somewhat, arguing that chemistry and physics are inseparable. The charge triggered by BMP, he explains, penetrates the cell and somehow rearranges the nucleotides. These are the basic chemical building blocks of deoxyribonucleic acid (DNA), the

chromosomal material determining heredity. He says that for a cell to change its level of specialization it must also change its genetic expression.

The power to peer deeper into the cell, to move closer to the secrets of the growth of bone, is the reward of inquiry. Such a search translates into keener understanding of how the body works and how, as well, to better repair it.

Biomechanically, the knee is perhaps the most wondrous yet vulnerable of the joints. Hanging free from the stable center of the body, the knee perpetually bends, glides and rotates in response to the stresses that motion puts on the joint. Four major ligaments bind the knee; supporting it are thirteen different muscles.

With every step, those ligaments and muscles contract, relax, twist and turn. Between the two long bones, the femur and the tibia, sit two crescent-shaped pieces of shock absorbent cartilage called the menisci. Spongy tissues, the menisci

Narrow incision and wide angle are allowed a surgeon, left, using an arthroscope. It affords a view of the total joint through a cut as small as a quarter-inch and permits the removal of little healthy tissue.

Like misbegotten moons, arthroscopic glimpses of the knee show the toll of arthritis. Jagged crystals reveal inflamed joint lining, at top. A thin equatorial line of blue marks eroded cartilage between bones, bottom.

are nearly avascular, meaning they are devoid of a blood supply. They can only absorb nutrients indirectly through fluid exchange and so have virtually no capacity for repair if damaged. When torn, usually by the simultaneous combination of excessive compression and rotation of the knee joint, this cartilage requires surgical removal. If it remains in the body it will begin to wear and initiate arthritis. Knee surgery has conventionally required cutting through layers of healthy tissue and removing much if not all of the cartilage. This is a procedure that has often produced considerable pain, expense and diminished chances for full recovery.

To See Inside the Knee

Arthroscopy has changed traditional knee surgery. First introduced by Japanese surgeon Kenji Takagi in 1918, the arthroscope's recent refinement permits the surgeon to operate on cartilage without opening up and exposing the knee. One day, perhaps, this tool might be used in more complex procedures such as the repair of torn ligaments or removal of bone fragments from a degenerated joint. The arthroscope has now shrunk to a nine-inch-long device that can fit through a quarter-inch incision. Designed like a telescope with half a dozen lenses aligned one above the other, the arthroscope bristles with optic fibers that throw light in the instrument's path like a lamp on a miner's hard hat.

What that light illumines, says Beverly Hills knee surgeon Jack Kriegsman, is the inner world of the knee. "Every time I look into it, I feel as though I'm taking a fantastic voyage," marvels Kriegsman. "I can see 100 percent of the joint. The arthroscope is small enough to get around behind all of the joint's structures." Inserting tiny knives through the incision, peering through the magnifying power of the arthroscope, the orthopedic surgeon is able to cut a minimum of healthy tissue and remove only that part of the cartilage that is torn, rather than the entire cartilage as is commonly done in open-knee surgery. The more cartilage that remains, the less prone the patient eventually is to osteoarthritis.

Kriegsman has followed up 360 arthroscopic operations which he has performed over the past

Bent by his heavy burden, Diego Rivera's The Flower Vendor *rises to his feet. Painted by the Mexican master in 1935, this oil and tempera on a wood panel shows more than the plight of Mexico's peasantry. Man's daily labor centers in the lower back, where the spine widens into the five lumbar vertebrae. When engaged with supporting muscles, they can bear enormous loads. A 100-pound burden held fifteen inches away from the body multiplies into 1,500 pounds of force on the lowest lumbar disk.*

six years and compared them with cases of open-knee surgery. While for the latter cases the average postoperative hospital stay lasted about six-and-a-half days, for Kriegsman's patients it was roughly a day. Arthroscopic surgery also saves the cost of the longer convalescence, adds Kriegsman, and significantly reduces — by about 75 percent — the need for postoperative pain-killing drugs. A new surgical procedure using the arthroscope also seems to hold out some hope for those suffering from osteoarthritis. The surgeon can roughen the surface of the bone, where the cartilage has been worn away, by manipulating a a diamond-encrusted burr through the arthroscope. Fibrocartilage, a type of scar cartilage, will then grow to replenish in some measure the cushioning function of the bone's surface.

Delicate Disk Surgery

Perhaps nowhere in the body is cartilage more integral to the functioning of the skeleton than in the back. From each of the thirty-three vertebrae, except for the top two, a pair of nerves branches and extends through the body, transmitting and receiving a range of sensations which they relay to the spinal cord and brain. Three membranous layers of tissue sheathe the spinal cord. It, in turn, threads through the rock-hard protective casing of the bony spinal column.

Between the vertebrae lie the disks, rubbery cartilaginous material that absorbs and disperses shock waves rolling up the spine. Like any other anatomical feature, disks vary from one person to the next. In some people, disks are more dense and better able to absorb shock.

A cross-sectional cut of a disk resembles a slice of onion. At its center lies a jellylike substance surrounded by concentric rings of fibrous tissue. Undue stress on the disk, coupled with a degenerated ring, can force the gel against the inner ring, and crack it. From there, the gel migrates outward, cracking successive rings in its path. Should the stress on the disk be severe enough, the gel will eventually break through the outer ring and pinch the nerve root leading from the vertebra. This is a "slipped" or herniated disk. It announces itself in the form of pain, numbness or muscle weakness in the extremities.

Back pain, a perennial problem, nonetheless seems more acute in modern, affluent societies. Lacking exercise, the network of muscles and ligaments supporting the edifice of the back falls into disuse and weakens. Without support, the disks become more prone to injury when stressed. Most particularly, such injury centers in the lower back, the lumbar segment of the spine, which bears the heaviest loads. Leaning forward and lifting weight, a man increases the load on his lumbar disks nearly 100 percent. Should he rotate his spine at the same time, he increases the pressure 400 percent. The result of such focused stress on the five lumbar vertebrae is that every year about twenty million Americans find themselves prey to back problems at the collective cost of $15 billion in lost wages and medical bills. Some 300,000 people require surgery yearly.

As in the case of the knee, recently developed tools and techniques make disk surgery a more precise and effective procedure. Today, the surgeon can peer through a microscope and part the tissue with forceps thinner than matchsticks. His scalpel now replaced by a tiny, pinlike instrument called a blunt dissector, the surgeon excises no more than the offending bulge of herniated disk pressing on the nerve. This operation is known as a microlumbar diskectomy. In the traditional disk operation, a standard laminectomy, the surgeon operates with the naked eye through a four-inch-long incision, removing much of the disk. In its place only scar tissue forms. The cushion of cartilage is lost forever.

Pioneered in 1973 by a Nevada neurosurgeon, Robert Williams, microlumbar diskectomy was greeted with skepticism by many in the medical world. They believed that the removal of such a small segment of the disk would lead to recurrent rupture of the cartilage. But after performing more than 1,000 such operations, Williams reports a "cure" rate of 98 percent. He says his procedure has shrunk the average hospital stay of his patients from nine days to three, since he began performing microscopic surgery instead of standard laminectomies. Blood loss in Williams's operation is also significantly reduced, a tenth less than what it is in the standard operation. Harold Goald, an Alexandria, Virginia neurosur-

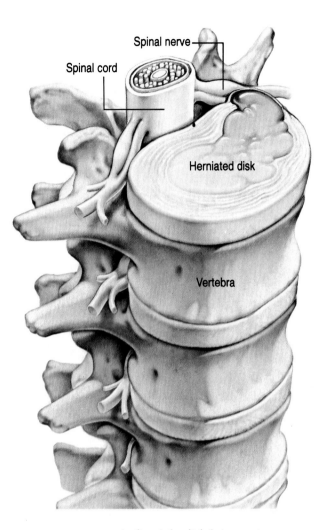

Spinal nerve

Spinal cord

Herniated disk

Vertebra

A slip of the disk brings pain or numbness to spinal nerves. Spongy cartilage layered through the spine, disks deaden shocks that resound through the vertebrae. If unduly stressed, their jellylike centers can bulge outward and come to press on a nerve. This is a "slipped" or herniated disk. A pervasive injury in affluent societies, slipped disks often result when muscles that guide and support the spine slacken from years of disuse.

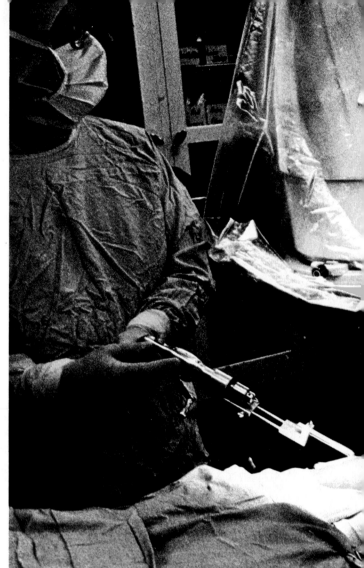

Healing a severe fracture, surgeons implant electrodes in a patient's leg. Below, current-carrying hormones trigger chemical messengers to the nucleus, providing a model of how electricity heals within the cell.

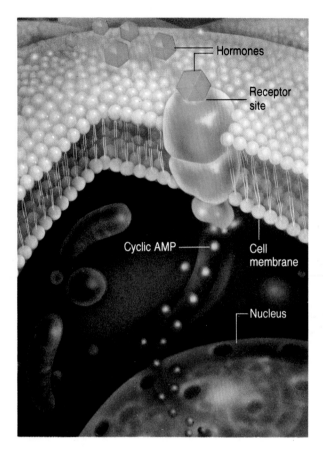

Hormones

Receptor site

Cyclic AMP

Cell membrane

Nucleus

geon, corroborates Williams's findings. He has performed about 1,000 microlumbar diskectomies and reports a similar success rate.

Only about 2,000 microlumbar diskectomies were performed in the United States in 1981, but Williams and Goald believe that this more exact surgical approach has the potential to replace the standard laminectomy — except in cases where the disks have severely degenerated or the spinal canal has narrowed. Goald estimates that the use of a microscope in surgery requires about six months' training to develop enhanced hand-eye coordination, an investment, he says, not every surgeon is willing to make. He adds that "like any new method, a period of time must pass before it is accepted."

The surgeon's urge to save spinal disks and knee cartilage is an effort to conserve both tissue and the life-giving forces that substance nurtures. One such force is electricity. It sweeps

through the fabric of the heart to kindle and sustain rhythm. It crosses the circuits of the brain, stirring awareness in its wake. It cradles its spark in bone's every cell. Any tissue with collagen in it — bone, tendon, cartilage and skin — sends an electric signal when mechanically stressed, injured or broken. No other tissues appear to have this characteristic.

In about 5 percent of all fractures, normal healing fails to occur. Soft tissue intercedes between the bone fragments. The broken bone may have an insufficient blood supply or number of regenerative cells. These are factors thought to contribute to nonhealing, or nonunion, fractures.

Over the past three decades scientists have been studying the physics of bone, with an eye toward enhancing its electrical potential in the mending of nonunion fractures. The technical products of that search are devices that stimulate bone growth electrically. The results have been

impressive thus far. In about 2,000 cases of non-union fracture, the two most commonly used devices — coils externally applied to the limb, and electrodes surgically implanted in the skin — boast a cure rate approaching 80 percent. Such an approach circumvents both the need for bone-graft surgery in the mending of nonunion fractures and the risks and costs that can ensue.

How electricity works in the healing of bone is a process just beginning to be understood. It appears that the currents generated by manmade devices are not strong enough to penetrate the cell wall. Yet the cellular membrane possesses the power to amplify the signal so that the nucleus of the cell can "hear" it and respond. Hormones carrying electrical potential seem able to interact with enzymes in the cell membrane. This then unleashes a "biochemical cascade," a torrent of molecules from a substance such as cyclic AMP (adenosine monophosphate). This biochemical messenger can pierce to the nucleus of the cell where it instructs the cell to activate its metabolism or initiate the cell division process, hastening the regrowth of bone.

Some scientists liken the healing of any wound or injury to a race between scar and healthy cells. The body tends to heal the connective tissues — bone, skin, cartilage — with their own kind of cells, and almost all other injured tissues with scar. Of all the internal organs, only the liver normally heals itself with its own tissue. Nerve tissue also displays a strong urge to heal itself with its own kind of cells but usually fails and mends with scar. With the aid of electricity, the body might one day be able to channel its recuperative powers and stimulate the growth of specific cells, be they unique to bone, liver, nerves or other tissues. Conversely, electricity may hold out the promise of limiting the growth of unwanted scar and even cancer cells.

Appendix

First Aid for Fractures

Every year, two million Americans suffer fractures — breaks or cracks in bone. Proper emergency care can ease pain and shock and help reduce risk of further damage to bones, nerves, blood vessels and soft tissues.

When parts of a broken bone puncture the surface of the skin, creating an open wound, the break is called an open fracture. In a closed fracture, the ends of the broken bone remain beneath the skin.

If you encounter someone who you suspect has fractured a bone, particularly in the neck or spine, do not move him or ask him to move the injured part. Make sure he is breathing properly. Give artificial respiration if necessary. If an open wound is visible, cut clothing away from it and cover it with a sterile or clean dressing. Apply pressure to curb severe bleeding, but be careful not to contaminate the wound. Then call for medical assistance.

If the victim felt or heard a bone snap or feels a grating sensation, suspect a fracture. Other symptoms are swelling, extreme pain at the site of the injury when slight pressure is

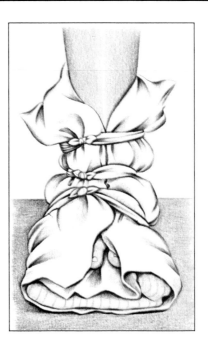

For fractures of the upper arm, place thin padding in the armpit. Gently move the arm to the victim's side, bending the forearm at a right angle across the chest. Place the splint on the outside of the upper arm and tie it above and below the break. Tie a sling around his neck to support the forearm. Wrap a towel or cloth around the splint, then tie it under the other arm.

Pillows or blankets can be used to splint a fractured foot or ankle. Remove the victim's sock and shoe and elevate his leg slightly. Slide the padding under the leg, making certain the material extends from above the calf to well below the heel. Tie the splint in place with bandages or strips of cloth. To support the foot, fold over the ends of the padding that extend below the heel.

Fractures of the upper leg, the largest bone in the body, are often accompanied by severe pain and deformity. A common sign is a shortened leg and a foot turned outward. First straighten the knee very carefully if it is bent and place padding between the legs. Using a stick, push long bandages through hollows under the victim's ankle, knee and lower back. Position them beneath applied, deformity of the injured part and unusual length or shape of a bone.

Before moving the victim, immobilize the injured part with a splint or bandage. If specialized splints are not available, use whatever materials are at hand. Splints can be improvised from sticks, poles, cardboard, umbrellas, brooms, blankets, tree limbs and rolled newspapers or magazines. If none of these is available, secure the injured area to another part of the body. A broken leg can be tied to the other leg with bandages.

Splints should extend beyond the joints above and below the suspected fracture. Never tie splints directly over a break. If the fracture is open, cover it with a sterile or clean dressing before applying the splint. Place padding such as towels, pillows or bedding between the splint and the victim's skin. Tie splints with rope, belts or strips of cloth. Make sure ties do not interfere with circulation. Loosen them if the victim's fingers or toes become swollen or turn blue, if he complains of numbness or tingling or if his pulse is faint.

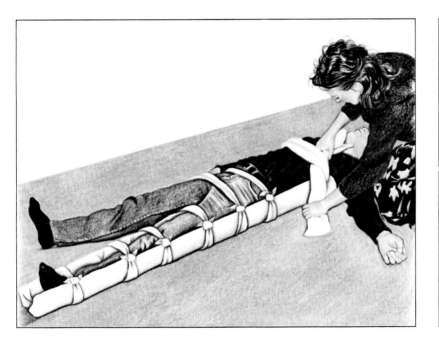

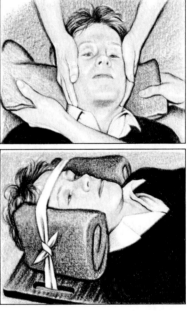

the armpit, lower back, pelvis, thigh, ankle and above and below the knee. Pad two boards, placing extra cushioning at the levels of the knee and ankle. Place one board along the outside of the leg, making sure it extends from the victim's armpit to below his heel, and another along the inside of the leg, reaching from the groin to below the heel. Tie bandages on the outer splint at three or four places. If boards or other long, rigid objects are not available, place rolled blankets or clothing between the legs and tie both legs together above and below the knee and around the thigh.

Splints for fractures of the lower leg can be made in the same way, with boards extending from above the knee to below the heel. Keep the victim's toe pointing upward.

Fractures of the neck and spine are the most critical. Call for medical help at once. Do not move the victim unless his life is in immediate danger. If you must move him, enlist the aid of several other people to minimize the risk of twisting his body. While waiting for medical assistance, place rolled blankets around his head and neck, then tie them in place very carefully.

A Guide to Back Care

Nearly everyone experiences back pain at one time or another. In rare instances, it signifies organic disease. But ninety-nine times in one hundred, backache is the result of physical strain or degenerative changes of the spine such as osteoarthritis or disk erosion. Whatever the cause, chronic pain and further damage can be prevented by learning how to use the body properly.

As a general rule, keep the back as straight as possible at all times. Back strain is more likely when the body assumes a position that increases any of the spine's natural curves. Poor posture can lead to strained ligaments, overlapping of joints and vertebral damage.

Bending and lifting: Always flex the knees and lower the hips rather than the waist when bending. This transfers most of

the burden from the delicate spine onto leg and thigh muscles. Squat, then pick up the object with the back slightly rounded, not arched. The entire body should move as a

single unit with shoulders and pelvis facing in the same direction. Movements should be slow and smooth. Avoid twisting, heaving or abrupt motions. Keep feet fairly far apart, with one ahead of the other for balance. Hold the object close to the body. If the load is very heavy, divide it into two lighter loads, one for each arm, or make two trips. Do not lift heavy objects higher than waist level. When possible, carry the load on other parts of the body such as the thigh or shoulders. To put a load down, either drop it or squat, placing one end or corner down first, then release the rest of it slowly.

Housework: Many household chores involve stressful bending and repetitive movements. Change chores and positions frequently. Kneel or sit when tucking in sheets to avoid stretching and bending. Use a long-handled mop or sponge when cleaning floors and bathtubs. The constant push-pull motion of vacuuming, mopping or raking can strain the

back and shoulders. Instead, stand sideways with feet apart and move the tool in a left-to-right motion rather than back and forth. Use short strokes and keep the tool close to the body. When working at a sink, sit on a stool or stand with one foot on a low footstool.

Shoveling snow is dangerous due to the strain of lifting a heavy weight at a relatively long distance from the body.

To shorten the distance, slide one hand down the shaft of the shovel. With very heavy loads, bend one knee and hold the handle against your knee or thigh. Swing each shovelful horizontally.

Sitting: People who sit for long periods of time can develop back problems from slouching or using the wrong chair. When sitting, each spinal disk must bear roughly twice as much pressure per square inch as it does when standing. Find a chair that enables you to maintain good posture effortlessly, with the spine nearly vertical. The back rest should support the spine, especially the lower, or lumbar, area. Place a cushion at the small of the back if it lacks support. The chair should be low enough so that both feet can be placed on the floor with knees bent at a comfortable angle. If the knees are not slightly higher than the hips, rest your feet on a low stool to help keep the lumbar area flat. Make sure the seat is deep enough for the knees to bend comfortably and firm enough so that your body does not sink into the chair. Place a board under the seat cushion if it is too soft. Arm rests also help reduce upper back strain by supporting the forearms.

Sit erect without arching the back or thrusting the head and neck forward. Keep buttocks tucked in and the lumbar area flat against the chair back. Shift positions frequently to

relieve pressure on the spine. If your job requires much sitting, take frequent breaks. Get up and walk around. If you spend much time reading or writing, use a tilted desk or a drawing

board propped up with books to prevent strain caused by bending the head forward. When typing, sit close enough to the typewriter so that you do not have to lift your entire arm to hit the keys. Arrange seat height so that typewriter keys are at waist level.

Car seats are frequently too low and soft. Place a firm cushion behind the small of the back and pull the seat forward so that the knees are slightly higher than the hips. You should not have to strain or lean forward to reach the pedals. Relax, shrug your shoulders and move your neck at stops to relieve tension.

Standing: Stand "tall," lifting the top of the head but keeping the chin in. The pelvis should be tucked under and the back kept as straight as possible. Keep knees relaxed rather than stiff. For added support and balance, keep feet at least as far apart as the shoulders, with toes pointing forward. Prolonged standing should be avoided. If you must stand while working, move

around as much as possible, lean against a table or shift body weight forward with knees bent to relieve pressure on the lower back.

Proper shoes are essential to good posture. Heels over two inches high exacerbate sway-back by throwing the body's weight so far forward that the back must be arched to maintain balance. Wear soft leather shoes with flexible soles.

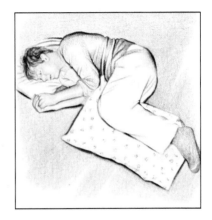

Sleeping: Choose a mattress that supports the body's weight evenly and keeps the spine properly aligned. If the mattress sags, place a plywood board three-quarters of an inch thick between the mattress and box spring. If the bed is so firm that it causes stiffness, place a blanket under the bottom sheet. Use a pillow just thick enough to keep the head and neck aligned with the spine. Sleeping on the stomach also increases swayback and should be avoided. Instead, sleep on your back with a pillow under your knees, or on your side with your knees bent and a pillow between the knees.

Glossary

acromegaly a disorder caused by excessive secretion of the pituitary growth hormone after maturity, causing enlargement of skeletal extremities such as the jaw, nose, hands and feet.

allograft transplanted tissue from one member of a species to another member.

anthropology the scientific study of the origins, development and behavior of man.

antibiotic a chemical substance derived from bacteria or fungi that destroys or impedes the growth of microorganisms; used to treat infectious diseases.

appendicular skeleton one of the two main anatomical categories of bones, consisting of all bones in the upper and lower extremities, including the shoulder and pelvic girdles.

arthroplasty the surgical reconstruction of a joint, either by repairing a damaged joint or replacing it with an artificial one.

arthroscope a telescopic instrument that permits the investigation of the interior of a joint.

arthroscopy the examination of the interior of a joint in diagnosis or surgery, using an arthroscope.

articular process one of the four smooth, slightly curved projections of a vertebra where it links with adjacent vertebrae.

articulation a joint; site where two or more bones connect. Articulations bind bones together and permit movement.

atlas the uppermost spinal vertebra that supports the head. Unlike other vertebrae, it has no main body.

Australopithecus a now extinct genus of hominids that lived during the Pleistocene epoch.

autograft tissue transplanted from one part of the body to another.

axial skeleton one of the two main anatomical categories of bones, consisting of the bones that form the body's upright axis: skull, vertebral column, sternum, ribs, hyoid and ear bones.

axis the second cervical vertebra; its body projects upward through the ring of the atlas above to form a pivot on which the head rotates.

biomechanics the study of the relationship between mechanical laws and human locomotion.

bipedalism the ability to stand and walk upright on two feet.

bone the rigid connective tissue that constitutes most of the human skeleton, consisting of collagen and numerous minerals; any specific skeletal component made from this substance.

bone graft the implantation or transplantation of bone from one part of the body to another or from one individual to another.

bone marrow a soft substance that fills the bone cavities. Red marrow, located in the cancellous tissue of the adult ribs, vertebrae, pelvis and skull bones, produces blood cells; yellow marrow, found in the center of long bones, consists of fatty material.

bone matrix the organic intercellular substance of bone composed of mucopolysaccharides and collagen fibers.

bone morphogenetic protein a protein that induces bone regeneration, useful in the surgical reconstruction of bone.

bunion inflammation of the bursa of the big toe.

bursa a small sac containing synovial fluid that helps ease friction between moving skeletal parts.

bursitis an inflammation of one or more bursae, often caused by excessive use of a joint, as in "tennis elbow."

calcaneus the heel bone; one of the seven tarsal bones of the foot.

calcitonin a hormone secreted by the thyroid gland that helps regulate calcium levels in the blood stream and stimulates bone formation.

calcium a mineral essential in the formation and growth of bone.

callus a mesh of new bone tissue that forms at the site of a bone fracture to be replaced ultimately with mature bone.

canaliculi microscopic channels radiating in all directions from, and connecting the lacunae of, a Haversian system; they provide a route for nutrient transport and waste removal via osteocytes.

cancellous bone one of the two basic types of bone, found at the ends of compact bones and in their hollow centers; also known as spongy bone because its latticelike structure contains many open spaces.

carpal bone one of the eight bones of the wrist, arranged in two rows of four each.

cartilage a firm, elastic connective tissue that forms most of the embryo's skeleton and is gradually converted to bone. It is found in the nose, ear, joints and at the ends of bones in the adult skeleton.

cartilaginous joint a joint in which bones are bound together with cartilage.

center of ossification an area where osteoblasts cluster to begin bone formation.

chondroblast a cell that produces cartilage.

clavicle the collarbone; a slim, S-shaped bone at the base of the neck which, along with the scapula, forms the shoulder girdle.

clubfoot a congenital deformity in which the foot is twisted out of shape; also called talipes.

coccyx the "tailbone;" a small, tapered bone beneath the sacrum at the base of the spinal column, comprised of four fused vertebrae.

collagen a tough, elastic protein found in bone, skin and all other connective tissue.

compact bone one of the two basic types of bone; dense bone tissue organized into closely packed concentric rings called Haversian systems.

compression the act of pressing together; an applied force or stress exerted on leg bones during walking or running.

coronal suture the line of union between the frontal bone and the two parietal bones across the front of the skull.

corticosteroid an organic compound produced by the adrenal gland, used in the treatment of rheumatoid arthritis.

costal cartilage one of twenty-four shafts of hyaline cartilage linking the seven pairs of true ribs to the sternum, the upper three false ribs with the lowest pair of true ribs and the lowest ribs with the abdominal muscles.

cranial index the ratio of the maximum width of the skull to its maximum length, multiplied by 100; sometimes used to classify skulls according to ethnic origin.

cranium the vault of the skull, consisting of eight bones which lodge and protect the brain.

Cro-Magnon man an early *Homo sapiens* subspecies with a large skull and a tall physique who lived in Europe during the late Paleolithic age.

cruciate ligaments two tough, thick ligaments crossing at the middle of the knee joint which prevent the joint from moving too far backward or forward.

diaphysis the shaft of a long bone.

dimethyl sulfoxide (DMSO) a colorless liquid used as an industrial solvent; also used experimentally as a topical analgesic and as an anti-inflammatory agent in the treatment of arthritis.

diode an electronic device containing an anode and a cathode, used in gait analysis to track skeletal motion.

dislocation the displacement of a bone from its normal position in a joint.

endochondral ossification the process of bone formation during which cartilage is converted into bone; most skeletal components are formed in this manner.

Eoanthropus dawsoni "Piltdown man;" fragments postulated to have been an early species of man but later discovered to have been a forgery; created from the skull of a man and the jaw of an ape.

epiphyseal plate a thin layer of cartilage between the end and the shaft of a long bone that enables the bone to increase in length.

epiphysis the end of a long bone.

evolution the continual and progressive change in the genetic makeup of organisms from one generation to the next.

facet a small rounded indentation on a bone at the point of juncture with another bone.

femur the thigh bone; the largest, longest and heaviest bone in the body, extending from the pelvis to the knee.

fibroblast a long, spindle-shaped connective tissue cell that forms fibrous tissues.

fibrous joint a joint where bones, attached by fibrous connective tissue, do not move.

fibula the smaller of the two bones of the lower leg, running parallel to and articulating with the other bone, the tibia.

fontanelle a soft, unossified space in an infant's skull.

foot binding the practice of deforming feet for aesthetic purposes, widely performed in China until the early twentieth century.

foramen magnum a large opening at the base of the skull through which the brainstem passes into the vertebral canal, where it becomes the spinal cord.

fossil the remains or imprint of an organism preserved in the earth's crust.

frontal bone one of the eight bones of the skull, forming the forehead and the top of the eye orbits.

gait analysis the study of the patterns of movement and the forces generated by the human body while walking; used in the diagnosis and treatment of joint disorders.

gout a form of arthritis thought to be hereditary and affecting mostly males, marked by the deposition of uric acid crystals in and around the joints.

Griffith crack a flaw due to an impurity in the molecular structure of bone crystals or metal.

Haversian canal a channel in the center of a Haversian system containing blood and lymph vessels, connective tissue and nerves.

Haversian system the basic structural unit of compact bone consisting of a Haversian canal and its surrounding concentric lamellae.

hematoma a blood clot that forms at the site of a bone fracture.

herniated disk an intervertebral disk whose gel-like nucleus has protruded through its outer layer, causing pressure on the spinal cord or spinal nerves; commonly called a slipped disk.

hip bone one of two heavy, earshaped bones which form the pelvic girdle. Each hip bone is made up of three bones (ilium, ischium and pubis) which fuse together during early adulthood.

hominid the family of primates including both living and extinct human and prehuman beings.

Homo erectus "Upright man;" the name given to the species represented by Java man and Peking man.

Homo habilis "Handy man;" an extinct species of the genus *Homo* that lived during the Pleistocene era; believed to be an ancestor to *Homo erectus.*

Homo sapiens modern man; the only living species of the hominid family.

humerus the long bone of the upper arm extending from the shoulder to the elbow.

hyoid a **U**-shaped bone at the base of the tongue.

ilium the largest and uppermost portion of the hip bone, situated at either side of the pelvic girdle.

incus a tiny anvil-shaped bone in the middle ear.

intervertebral disk a strong fibrocartilaginous pad nestled between spinal vertebrae which helps absorb shock and prevents the vertebrae from grating against each other.

intramembranous ossification the formation of bone within a membrane rather than within cartilage.

ischium the strongest and lowermost portion of the hip bone.

Java man an apelike prehuman with heavy brow ridges and a small brain capacity that lived during the middle Pleistocene epoch; also known as *Pithecanthropus* or *Homo erectus.*

joint an articulation; the site where two or more bones meet, binding the bones firmly together and permitting movement between them.

joint capsule a sleevelike extension of the periosteum lined with a fluid-secreting membrane that encases a synovial joint.

joint cavity a space in the core of a synovial joint that permits free movement between bones.

kyphosis an increased convexity of the spine, most commonly seen in the thoracic vertebrae.

lacuna a small cavity containing bone cells that creates a flow of nutrients through interconnecting small channels (the canaliculi) within the lamellae of a Haversian system.

lambdoid suture the line of union between the occipital bone and the two parietal bones across the back of the skull.

lamella one of the thin plates of bone arranged in concentric layers surrounding a Haversian canal.

laminae broad plates of bone flaring out from either side of a vertebral pedicle and fusing together to form a circular opening for the spinal cord.

laminectomy the surgical removal of a portion of the posterior arch of a vertebra and part of the intervertebral disk.

ligament a band of fibrous tissue that connects bone or cartilage and helps strengthen and support joints.

lordosis an exaggerated concave curve in the lumbar area of the spine.

lotus foot a condition created by foot binding, popular in China until the 1930s, in which the curve of the instep was steeply exaggerated and the four smaller toes curled under the ball of the foot.

lumbar the region of the back between the ribs and the hip bones.

malleus the largest of the three bones of the middle ear. Shaped like a hammer, it passes along vibrations from the eardrum to the other two bones, the incus and the stapes.

Mammalia a class of warm-blooded vertebrates including man and all others that have a body covering of hair, three small bones in the middle ear and, in females, mammary glands.

mandible the lower jawbone; the largest and strongest facial bone.

maxilla one of the two main facial bones which constitute the upper jaw and form parts of the base of the eye orbits, the roof of the mouth and the floor and sides of the nose.

medullary cavity a hollow space in the diaphysis of a long bone that contains yellow marrow.

meniscus a crescent-shaped wedge of fibrocartilage found in some synovial joints having a slippery surface that reduces friction and absorbs shock.

mesenchyme a part of the mesoderm, or middle layer, of an embryo from which the skeletal tissues, connective tissues, blood vessels and lymphatic system develop.

metacarpal bone one of the five bones that radiate from the wrist to form the palm of the hand.

metatarsal bone one of the five slender bones that articulate with the bones of the ankle and toes to form the arch of the foot.

microlumbar diskectomy the surgical removal of all or part of a herniated intervertebral disk.

mineralization the process of fossilization in which the organic remains of an organism are converted into an inorganic mineral substance.

mucopolysaccharide a cement-substance secreted by osteoblasts that unites with collagen fibers to form the bone matrix.

nasal septum a vertical partition made of bone and cartilage that separates the right and left nasal openings.

natural selection a process that favors the survival of and propagation of those individuals and their offspring best adapted to their environment.

Neandertal man an extinct subspecies of *Homo sapiens* that inhabited Europe during the late Pleistocene age and became extinct 35,000 years ago; distinguished by its low, broad skull, prominent brow ridges and short limbs.

nucleotide a basic structural component of deoxyribonucleic acid (DNA), composed of sugar, phosphoric acid and a nitrogenous base.

obturator foramen a large opening at the base of the pelvis between the pubis and the ischium.

occipital bone the bone that forms the lower posterior part of the skull.

occipitomastoid suture an extension of the lambdoid suture between the occipital bone and the downward projection of the temporal bone, at the base of the skull.

orbit a deep, bony cavity in the skull that holds the eyeball.

orthopedics a branch of surgery that treats skeletal disorders.

ossification the formation of bone; the conversion of hyaline cartilage or fibrous membrane into bone.

osteoarthritis a chronic degenerative joint disease, most often seen in the elderly, marked by the breakdown of cartilage and the formation of bone spurs.

osteoblast a bone-forming cell.

osteoclast a bone-destroying cell.

osteocyte a mature osteoblast lodged in the lacunae of a Haversian system.

osteoinduction the conversion of fibroblasts into cartilage-producing cells.

osteomalacia the softening of bones due to a lack of vitamin D, calcium, phosphorus or of all three.

osteoporosis a disorder often seen in the elderly that causes bones to become thin, soft and weak.

Paget's disease a disorder marked by increased bone resorption followed by excessive regrowth, resulting in bone expansion and weakening.

paleoanthropology a branch of anthropology that studies manlike creatures more primitive than *Homo sapiens*.

pannus an overgrowth of connective tissue inside a joint that corrodes cartilage; often seen in rheumatoid arthritis.

parathyroid hormone (PTH) a substance secreted by the parathyroid gland that stimulates the breakdown of bone.

patella the kneecap; a small, flat, movable bone protecting the knee joint.

pedicle a short shaft of bone rising from either side of a vertebra to form the vertebral arch.

Peking man an extinct humanlike species of *Homo erectus* living during the Pleistocene epoch whose fossil remains were found near Peking; distinguished by its thick cranium and heavy brow ridges.

pelvis a basinlike structure that supports the trunk and attaches the legs to the rest of the skeleton. It is formed by the two hip bones, the sacrum and the coccyx.

perichondrium a dense fibrous layer covering cartilage except at the joints.

periosteum a dense, tough fibrous membrane laced with blood vessels that covers most surfaces of all bones, essential for bone growth and repair.

phalange a bone in the finger or toe.

phosphorus a mineral involved in many metabolic processes and a major component of bone.

piezoelectricity an electrical current generated by the interaction between crystals and collagen fibers in bone.

Pleistocene epoch a geologic epoch of the Quaternary period, also called the Ice Age, beginning two million years ago and lasting one-and-a-half million years.

Primates the highest order of mammals, including man, apes and monkeys; distinguished by a relatively large cerebral cortex, nails rather than claws and stereoscopic vision.

prostaglandins a group of chemically related fatty acids that regulates certain body functions including blood pressure, body temperature and acid secretions of the stomach; often found in overabundance among rheumatoid arthritis patients.

pubic symphysis a cartilaginous joint at the juncture of the two pelvic bones.

pubis the anterior portion of each hip bone; the pubic bones join at the front to form the pelvic arch.

purine a crystalline compound that yields uric acid upon metabolism.

pyknodysostosis a rare disease causing dwarfism or incomplete bone growth.

radius the shorter of the two lower arm bones, located on the inside of the arm.

resorption the destruction of bone by osteoclasts.

rheumatoid arthritis a chronic inflammation of the synovial membrane lining a joint, often causing pain, disability and deformity.

ribs twelve pairs of curved bones extending from the thoracic vertebrae toward the sternum. The upper seven pairs ("true ribs") connect with the sternum directly; the lower five pairs ("false ribs") are indirectly linked. The lowest two pairs of false ribs, called "floating ribs," make no connection with the sternum.

rickets a disease of children that prevents normal bone formation, caused by a deficiency of vitamin D, resulting in soft and often deformed bones.

rondelle a bone carved from a trephined skull, often used as an amulet.

sacrum a triangular bone, formed by five fused vertebrae, that forms a wedge between the two hip bones.

sagittal suture the vertical line of union between the two parietal bones of the skull, connecting the coronal and the lambdoid sutures.

scamnum a bench designed by Hippocrates equipped with crankshafts and levers, and used to set fractures and dislocations.

scapula the shoulder blade; a large, triangular bone in the back of the shoulder. The scapula and clavicle together form the shoulder girdle.

sciatic notch a groove in the posterior border of the hip bone where the ilium and ischium unite.

scoliosis a lateral curvature of the spine.

shear the simultaneous application of forces from opposite directions.

shoulder girdle a flexible structure composed of the clavicles and scapulae which connects the arms to the axial skeleton.

sinus an air cavity in some skull bones, either frontal, ethmoid, maxillary or sphenoid, which links with the nasal cavity and adds resonance to the voice.

skeleton the internal framework of the body composed of bone and cartilage. It protects and supports internal body parts, allows movement and acts as a reservoir for certain minerals.

skull a skeletal case enclosing the brain, consisting of the bones of the cranium and the face.

sphenoid bone a wedge-shaped bone at the base of the cranium that anchors most other cranial bones and forms part of the sides of the eye orbits.

spinal column the S-shaped backbone extending from the cranium to the coccyx, formed by thirty-three vertebrae and interconnecting disks. It holds the trunk and head erect and houses the delicate spinal cord within its inner channel.

spinous process a winglike extension projecting downward from the arch of a vertebra that attaches to muscle.

splint a rigid device used to immobilize a fractured or injured part.

stapes the innermost bone of the middle ear, resembling a stirrup.

sternum the breastbone; a daggerlike structure to which most of the ribs are attached.

suture a fibrous immovable joint uniting the bones of the skull.

synovial fluid a lubricating substance secreted by the membrane of a synovial joint.

synovial joint the most mobile of the three basic types of joints, having a moist membrane that lines the inner surface of its joint capsule.

synovial membrane a moist sheet of tissue that secretes a lubricating fluid.

talus an ankle bone; the highest of the tarsal bones of the ankle.

tarsal bone one of the seven short bones of the ankle.

temporal bone one of the two bones forming the lower sides and part of the floor of the skull.

tendon a fibrous cord of connective tissue that attaches muscle to bone.

tension a type of force or stress exerted on bones when they are stretched.

thorax the chest; the bony cage formed by the ribs, protecting the heart and the lungs.

thyroxin a hormone secreted by the thyroid gland that regulates metabolism and bone growth.

tibia the shinbone; the larger of the two bones of the lower leg.

trabeculae thin slabs of connective tissue in cancellous bone, joined into a scaffoldlike structure along lines of stress to give bones added strength.

traction pulling; a treatment for bone fractures and dislocations using weights.

transverse process a winglike projection on either side of a vertebra which attaches to muscle.

trephination a procedure in which a portion of bone from the skull is removed with a sawlike instrument or sharpened stone; used among primitive cultures possibly to free a patient from demons or to relieve headaches.

ulna the longer of the two bones of the forearm, located on the side opposite the thumb.

uric acid a crystalline compound, the end product of purine metabolism in man.

vertebra one of the thirty-three small bones that form the spinal column. Each one consists of a thick oval body, an arch forming a channel for the spinal cord and numerous projections for linkage to muscles and adjacent vertebrae. The seven vertebrae of the neck are called cervical; twelve thoracic vertebrae anchor the upper back; five lumbar vertebrae make up the lower back; by adulthood, five sacral vertebrae fuse to form the sacrum and four coccygeal vertebrae fuse to form the coccyx.

Vitallium an alloy of cobalt, chromium and molybdenum used in artificial joints and surgical appliances.

vitamin D an organic substance required for normal bone growth that increases the rate of absorption of calcium and phosphorus from the intestine.

Volkmann's canal a tiny channel through which blood vessels and nerves from the periosteum penetrate compact bone in order to transport nutrients between Haversian systems.

Wolff's law a principle holding that the shape and structure of a bone is dependent on the forces acting on it.

Zinjanthropus boisei "East African man," also known as *Australopithecus boisei;* a semierect primate subgenus that lived during the Pleistocene epoch.

zygomatic bone the cheekbone, situated below each eye, forming the outer wall and part of the floor of the eye orbit.

Illustration Credits

Introduction
Photograph Agfa-Gevaert made on Agfa-contour film.

An Ageless Form
8, Pompeii/Skull Mosaic, SCALA/ Editorial Photocolor Archives. 10, The Mansell Collection. 11, (top) The Granger Collection, New York, NY (bottom) Lee Boltin. 12, *Caw Wacham* by Paul Kane, all rights reserved. Courtesy of the Montreal Museum of Fine Arts. 13, (top) Courtesy of the Essex Institute, Salem, MA (bottom) **Graziella Becker**. 14, Collection Muse de l' Homme. 15, (top) *The Extraction of the Stone of Madness* by Hieronymus Bosch; all rights reserved © Museo del Prado, Madrid (bottom) Lee Boltin. 16, From the Library of the New York Academy of Medicine. 17, Photo Giraudon, Paris. 18 and 19, The Bettmann Archive, New York, NY. 20, The Granger Collection, New York, NY. 21, Canali, Rome. 22, National Library of Medicine with help from Lucinda Keister. 23, The Bettmann Archive, New York, NY. 24, Ann Ronan Picture Library. 25, Royal Library, Windsor Castle; reproduced by gracious permission of Her Royal Majesty Queen Elizabeth II. 26, Biblioth que Nationale, Paris. 27, The Bettmann Archive, New York, NY. 28, **Thomas B. Allen**. 29, (both) National Library of Medicine. 30, Courtesy of the Rijksmuseum-Stichting, Amsterdam. 31, Editorial Photocolor Archive. 32, From the Library of the New York Academy of Medicine. 33, The Bettmann Archive, New York, NY. 34, **Thomas B. Allen**. 35, The Bettmann Archive, New York, NY.

The Flexible Framework
36, *Nude Descending a Staircase, Number 2, 1912* by Marcel Duchamp, Philadelphia Museum of Art: The Louise and Walter Arensberg Collection. 38, (top) Courtesy of Field Museum of Natural History, Chicago, IL (bottom) Animals Animals/ L. L. T. Rhodes. 39, **Eugenia Walda**, with the help of Dr. George E. Watson, Smithsonian Institution. 40, Vision International. 41, **Pat Kenny**. 42, (left) Manfred Kage/Peter Arnold, Inc. (right) **Pat Kenny**. 43, The Bettmann Archive, New York, NY. 44, Sonia Halliday & Laura Lushington. 45, **Jane Gordon**. 46, National Library of Medicine. 47, Jeffrey A. Zeisler for the Laboratory of Arthur M. Siegelman/FPG. 48, National Library of Medicine. Foldout (outside) Librarian, University of Glasgow (inside) **Dan Osyczka and Ken Goldammer**. 49, **Ray Srugis**. 50, Lennart Nilsson from his book *Behold Man*, published in the U.S. by Little, Brown & Co., Boston, MA. 51, **Susan Sanford**. 52, Douglass Baglin/ FPG. 53, Derek Ellis, U.K. 54, **Susan Sanford**. 55, (top) Photri/W. Cannon (bottom) Sonia Halliday & Laura Lushington.

Matters of Bone
56, Michael Abbey/Photo Researchers, Inc. 58, National Library of Medicine. 59, (top) Photograph by Carolina Biological Supply Co. (bottom) M. I. Walker/Photo Researchers, Inc. 60, **Jennifer Arnold**. 61, Neil Leifer/*Time* magazine. 62, **David Mascaro**. 63, Biology Media/Photo Researchers, Inc. 64, National Library of Medicine. 65, Taurus Photos/Alfred Owczarzak. 66, © Lennart Nilsson from his book *Behold Man*, published in the U.S. by Little, Brown & Co., Boston, MA. 67, (left) John Watney Picture Library (right) From Milch, *Journal of Bone and Joint Surgery*, vol. 22, 1940. 68, Editorial Photocolor Archives. 69, **Thomas B. Allen**. 70, (left) **Lewis Calver** (right) Jonathon T. Wright/Bruce Coleman, Inc. 71, **Joyce Hurwitz**. 72, (left) National Library of Medicine (right) © London Scientific Fotos. 74, (left) Ed Musy (right) Lester V. Bergman & Associates, Cold Spring, NY. 75, (left) Lester V. Bergman & Associates, Cold Spring, NY (right) Ed Musy. 76, (left) Ed Musy (right) Lester V. Bergman & Associates, Cold Spring, NY. 77, (left) Lester V. Bergman & Associates, Cold Spring, NY (right) Ed Musy. 78, Lester V. Bergman & Associates, Cold Spring, NY. 79, (left) Richard Pilling/Focus on Sports (right) Mickey Palmer/Focus on Sports.

Growth and Renewal
80, *Anatomical Painting* by Pavel Tchelitchew, Collection of Whitney Museum of American Art. Gift of Lincoln Kirstein. 82, © Biology Media/ Photo Researchers, Inc. 83, The Bettmann Archive, New York, NY. 84, (top) **Scott Barrows** (bottom) Biophoto Associates. 85, Dr. George P. Bogumill, Georgetown University School of Medicine, Washington, DC. 86, **Thomas B. Allen**. 87, From *Tissues and Organs: A Text-Atlas of Scanning Electron Microscopy* by Richard G. Kessel and Randy H. Kardon. W. H. Freeman and Company. Copyright © 1979. 88, Bruno Barbey/ Magnum Photos, Inc. 90, (left) The Granger Collection, New York, NY (right) J. L. Angel, Physical Anthropology, Smithsonian Institution. 91, (top left) Copyright The Frick Collection, New York, NY (top and bottom right) J. L. Angel, Physical Anthropology, Smithsonian Institution (bottom left) SCALA/ Editorial Photocolor Archives. 93, (left) Dr. George P. Bogumill, Georgetown University School of Medicine, Washington, DC (right) **Graziella Becker**. 94, FPG/Arthur M. Siegelman. 95, Dr. George P. Bogumill, Georgetown University School of Medicine, Washington, DC. 97, National Gallery of Art; Samuel H. Kress Collection. 98, Copyright G. D. Hackett, New York, NY. 99, **Thomas B. Allen**. 101, Dr. George P. Bogumill, Georgetown University School of Medicine, Washington, DC.

A Measure of Immortality
102, *Hamlet et Horatio*, Eugene DelaCroix, Photo Giraudon, Paris. 104, J. E. Harris, University of Michigan, Ann Arbor, MI. 105, (left) David Brill © National Geographic Society, Washington, DC (right) M. H. Wolpoff. 106, J. L. Angel, Physical Anthropology, Smithsonian Institution. 107, **Thomas B. Allen**. 108, **Jane Gordon**. 109, T. Dale Stewart, From *Essentials of Forensic Anthropology* 1st Ed., 1979, Courtesy of Charles C. Thomas, publisher, Springfield, IL. 110, Lee Boltin. 111, T. Dale Stewart, Physical Anthropology, Smithsonian Institution. 112, Lee Boltin. 113, David Brill © National Geographic Society, Washington, DC. 114, John Reader, Courtesy of Rheinisches Landesmuseum, Bonn. 115, © Field Museum of Natural History, Frederick Blaschke. 116 & 117, Jay Matternes, Courtesy of *Science 81*. 118-119, **Joyce Hurwitz** with help from Raymond Rye, Smithsonian Institution. 119, (bottom) Mary Evans Picture Library. 120, John Reader, Courtesy of The Dubois Collection, Rijksmuseum von Natuurlijke Historie, Leiden. 121, The Illustrated London News Picture Library. 122, John Reader, Courtesy of the Department of Anatomy, University of the Witwatersrand Medical School, Johannesburg. 123, **Thomas B. Allen**. 124, Emory Kristof © National Geographic Society, Washington, DC. 125, John Reader, Courtesy of the National Museum of Tanzania and M. D. Leakey. 126, John Reader, Courtesy of the Kenya National Museum. 127, Marion Kaplan, Camera 5. 128, John Reader, Courtesy of Dr. D. C. Johanson, Cleveland Museum of Natural History on behalf of the National Museum of Ethiopia. 129, David Brill, © National Geographic Society, Washington, DC. 130 & 131 (left), Jay Matternes, Courtesy of the National Geographic Society, Washington, DC. 131, (right) John Reader © National Geographic Society, Washington, DC.

Form and Future
132, *The Skull of Zurbaran* by Salvadore Dali, Hirshhorn Museum and Sculpture Garden, Smithsonian Institution. Photo by John Tennant. 135, **Robert J. Demerest**, courtesy of Eli Lilly and Co. 136, **Carol Donner**. 137, **Scott Barrows**. 138, Smithsonian Institution Photo No. 73-7898. 139, Dan McCoy/Rainbow. 140, **Thomas B. Allen**. 141, Lester Kalisher, M.D.; St. Barnabas Medical Center, Livingston, NJ. 142 and 143, Joe McNally/Camera 5. 145, **Thomas B. Allen**. 146, R. W. Jackson, M.D., M.S., F.R.C.S. (C); Associate Professor, University of Toronto. 147, (both) Jack Kriegsman, M.D.; Beverly Hills, CA. 148, *The Flower Vendor* by Diego Rivera, San Francisco Museum of Modern Art. Gift of Albert M. Bender in memory of Caroline Walter. 149, **Ray Srugis**. 150, **Carol Donner**. 150-51, Zimmer/ProClinica.

Appendix, 152-155, **Donald Gates**.

Index

Upright man, *see Homo erectus*
uranium, 115
uric acid, 134
Urist, Marshall, 144, 145, 146

V
Van Deventer, Henrick, 31
Venable, C. S., 138
Venel, Jean André, 32
vertebra, 24, 30, 41, **41**, 43, 67, 68, 71, 72,
    73, 76, 82, 89, 93, 106, 148, 149, **149**
  atlas, 42
  axis, 42, 43
  body, 41, 42
  cervical, 41, 43
  lumbar, 41
  thoracic, 41, 43, 45

vertebral column, 87
Vesalius, Andreas, 29, 30, 37, 45, 46, 51
villi, 87
Virchow, Rudolph, 35
virus, 134, 136
vitamin, 95
  A, 84, 96
  C, 84, 96
  D, 84, 95-96
Volkmann's canal, **62**, 65

W
Watson-Farrar, John, 141
weightlessness, 89
White, Charles, 33
Williams, Robert, 149-50

Wolff, Julius, 89
Wood, Bernard, 126, 129
Woodward, Arthur Smith, *see* Smith
    Woodward, Arthur
wrist, *see* arm

X
X-ray, 7, 28, 35, 53, 78, 88, 95, 100, 101,
    104, 107

Z
zinc, 58
Zinj, 125-26, 129, 130
Zinjanthropus boisei, 126
zircon, 115
zygomatic bone, 51